essential atlas
of anatomy

BARRON'S

Original title of the book in Spanish: *Atlas de Anatomía*.
© Copyright 2000 by Parramón Ediciones, S.A.—World Rights
© Copyright of English-language edition 2001 by Parramón Ediciones, S.A.
Published by Parramón Ediciones, S.A., Barcelona, Spain

Author: Parramón's Editorial Team
Scientific Advisor: Dr. Adolfo Cassan
Illustrator: Parramón's Editorial Team
Designer: Toni Inglés Studio

English-language edition for the United States, its territories and dependencies, and Canada published 2001
by Barron's Educational Series, Inc.

All inquiries should be addressed to:
Barron's Educational Series, Inc.
250 Wireless Boulevard
Hauppauge, NY 11788
http://www.barronseduc.com

International Standard Book No.: 0-7641-1833-1

Library of Congress Catalog Card No.: 2001135074

Printed in Spain
9 8 7 6 5

FOREWORD

This *Essential Atlas of Anatomy* offers readers a magnificent opportunity to learn more about the human body and the structure of the different components of the human organism. Indeed, it offers a highly useful tool for getting a closer understanding of the wonder of the human body, an organism that is often compared to a complex machine. In reality, the human body is much more than that. It is infinitely more intricate than any machine that anyone has ever designed up to now and, one can state with absolute certainty, than any machine that will ever be designed in the future.

The different chapters of this work represent a complex summary of the human anatomy. It consists of multiple sections and many illustrations that are diagrammatic but rigorous. They show the main characteristics of each and every one of the different apparatuses and systems of the human body. These illustrations, which represent the central core of this book, are complemented by explanations and notes. These aid in understanding the main anatomic and physiological concepts. The book also has an alphabetical index that permits the reader to find any point of interest with ease.

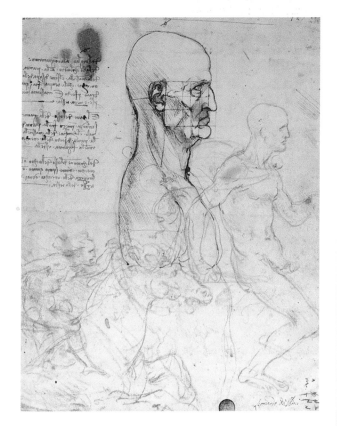

In deciding to publish this *Essential Atlas of Anatomy*, our objectives have been to create a practical, didactic work that is useful and accessible while, at the same time, being scientifically rigorous, clear, and easy to use. We hope the reader will think we have fulfilled our goal.

CONTENTS

Introduction **6**

The cell **10**
Components of the human cell 10

The human body **12**
Male anatomy 12
Female anatomy 13

The locomotive system **14**
The skeletal system 14
    Bone tissue 14
    Formation and growth of bone 14
    Types of bone 15
    Osseous blood supply 15
    Bone fractures 15
    The skeleton (anterior view) 16
    The skeleton (dorsal view) 17
    The skull bones 18
    Vertebral column 19
    Skeleton of the upper limb 20
    Skeleton of the lower limb 21
Joints 22
    Types of joints 22
    Dislocations 22
    Knee joint 23
    Shoulder joint 23
    Hip bone 23
    Meniscus 23
The muscular system 24
    Muscle structure 24
    Muscle shapes 24
    Muscles of the human body (anterior view) 25
    Muscles of the human body (posterior view) 26
    Muscles of the human head 27
    Muscles of the upper limb 28
    Muscles of the lower limb 29

The digestive system **30**
The digestive process 30
Buccal cavity 31
    Section of a tooth (molar) 31
    Deciduous teeth 31
    Permanent teeth 31
Esophagus 32

Swallowing 32
Stomach 33
Small intestine 34
Pancreas 35
Liver 36
Gallbladder and bile ducts 36
Large intestine 37

The respiratory system **38**
Organs of the respiratory apparatus 38
Mechanism of breathing 38
Bones and cartilage of the nasal pyramid 39
Nose 39
Lateral section of pharynx 39
Larynx and trachea 40
Bronchial tree 41
The lungs 41

The circulatory system **42**
Diagram of the circulatory system 42
Heart 43
Cardiac valves 44
Coronary blood vessels 45
Coronary system of electric impulses 45
Cardiac cycle 45
Main arteries of the organism 46
Main veins of the organism 47

Blood **48**
The composition of blood 48
Blood cells 48
Bone marrow 49
Spleen 49

Lymph **50**
Relationship between the lymphatic
system and the circulatory system 50
Lymphatic capillary 50
Lymphatic vessels 51
Schematic representation of the
    lymphatic system 51
Lymph node 51

The nervous system **52**
Components of the nervous system 52

Structure of a neuron 52
Types of neurons 52
The brain 53
Cerebrum 54
Meninges 55
Areas of the brain 55
Vertebral column and spinal cord 56
Peripheral nervous system 57

The senses **58**
Sight 58
    The eye 58
    Projection of images onto the retina 58
    Lacrimal apparatus 59
    Conjunctiva 60
    Optic nerve 60
    Visual pathways 61
    Principal defects of sight
       and methods of correcting them 61
Hearing 62
    External ear 62
    Ossicles of the middle ear 62
    The auditory process 63
    Labyrinth 63
Smell 64
Taste 65
    The tongue 65
    Areas of perception of
       different tastes 65
Touch 66
    Section of the skin 66
    Sensory receptors 66
    Reflex action to pain stimulus 67
    Hair follicle and root 67
    Nail 67

The urinary system **68**
Components of the urinary system 68
Renal circulation 68
Kidneys and blood supply 69
Section of a kidney 69
Blood vessels of the kidney 70
Nephron 70
Urinary bladder 71
Urethra 71

The reproductive system **72**
Male genital organs 72
    Penis 73
    Testicle and epididymis 73
    Prostate 73
Female genital organs 74
    Vagina 75
    Ovary and ovarian follicle 76
    Breasts 76
Menstrual cycle 77

Human reproduction **78**
Coitus 78
Union of an ovum and a spermatozoon 78
Fertilization and implantation 79
Gestation 80
    Development of the embryo 80
    Development of the fetus 81
Placenta 82
Abdomen of a pregnant woman 83
Birth 84
    Positioning of the fetus 84
    Fetal positions 84
    The delivery process 85

The endocrine system **86**
Glands of the endocrine system 86
    Hypothalamus and pituitary gland 86
    Functions of the hypothalamus 87
    Hormonal secretion of the
       pituitary gland 87
    Thyroid gland 88
    Parathyroid glands 88
    Suprarenal glands 89

The immune system **90**
Organs of the immune system 90
    Location of the thymus 90
    Mechanism of nonspecific
       immunity 91
Active immunization: vaccination 91

Alphabetical index of subjects **92**

INTRODUCTION

THE FIRST STUDIES OF HUMAN ANATOMY

The term *anatomy* comes from the Greek word meaning "dissection." It is used to describe both the structure of living beings as well as the science dedicated to studying them. Anatomy is an ancient science. In terms of human anatomy, though, it began to take real shape in the middle of the sixteenth century when scientists began to dissect bodies methodically in order to study their constitution properly. Before this period, the practice was forbidden, being categorically rejected by ethical views and reigning religious beliefs of the time. All knowledge that previously existed concerning the structure and working of the human body came, above all, from observations made on animals, and they were imprecise and often simply wrong.

The Belgian doctor Andrés **Vesalio** (1514–1564), professor of anatomy at such prestigious universities as Lovaina, Padua, and Bologne, dared to disobey the moral constraints of his time. He began to combine theoretical explanations in his classes with practical demonstrations based on the dissection of human bodies. In 1543, he published his findings in a book entitled *De humani corporis fabrica* (*The Structure of the Human Body*), which contained more than three hundred meticulous anatomic engravings. This work caused a great uproar among his fellow scientists because he openly contradicted the theories that had been accepted up until then.

It represented an enormous step toward a precise understanding of the human body. For its illustrious author, though, it brought severe repercussions. In 1561, while he was living in Spain and offering his services to the court of Philip II, he was put on trial by the Inquisition because of his daring and condemned to death. Later, though, his sentence was commuted to making a pilgrimage to the Holy Land. On the way back, he was killed in a shipwreck.

With the passing of time, anatomic studies became normal practice. **Findings made by simple observation** through dissection were added to findings made with increasingly modern **techniques**. In this way, scientists were able to understand with much greater accuracy how the human body is constituted and what functions each of its components fulfills. Therefore, the apparatuses and systems of the body could be better understood. This helped to identify the different tissues that make up the organs and the nature of their elemental components, the cells. Since the human body is extremely **complex**, it has millions and millions of components that combine in an extremely intricate way.

CELLS AND TISSUE

In essence, the human body is made up of an enormous number of **cells**, which are the basic units of all living beings. In fact, it has been calculated that an adult human being has more than two

The cell

The human body

The locomotive system

The digestive system

The respiratory system

The circulatory system

Blood

Lymph

The nervous system

The senses

The urinary system

The reproductive system

Human reproduction

The endocrine system

The immune system

Alphabetical index

hundred billion cells. Each is equipped with similar elements, although these elements take different forms and are designed to fulfill different specific functions.

These different cells are not arranged at random. Instead, they are grouped according to their characteristics. They are often combined with inert materials such as mineral salts or fibers that they themselves produce to form **tissue**. The human body has four basic types of tissue, each of which has a different task to perform.

• **Epithelial tissues** are formed of very similar cells that are closely linked to one another. Their most important function is both as a covering and for secretion. These tissues cover the external surface and internal cavities of the body. They also make up the glandular structures that secrete various substances and transport them either to the exterior of the body or into its interior, such as to the bloodstream.

• **Connective tissues** are composed of different type of cells, between which are overlaid substances of variable consistency. These include protein-based fibers, whose most important function is to provide support to body structures. In fact, there are different types of connective tissues. Loose connective tissue is found throughout the organism and has a fundamental role in the nutrition of all tissues because it allows blood vessels to pass through it.

Dense connective tissue is very strong and makes up tendons and ligaments. There are also other specialized types of connective tissue with specific properties, such as adipose tissue, cartilaginous tissue, bone tissue, blood tissue, and lymphoid tissue.

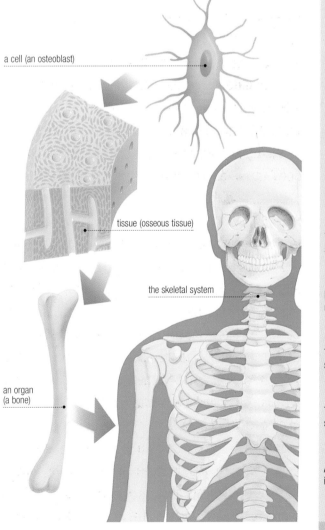

a cell (an osteoblast)

tissue (osseous tissue)

the skeletal system

an organ (a bone)

• **Muscle tissues** are made up of long cells that are capable of contracting when given the appropriate stimulus and then recovering their initial dimensions. Their task is to provide mobility for the body and its internal structures.

• **Nervous tissues** consist of very special cells, the neurons. These are capable both of receiving and generating stimuli and of transmitting information in the form of electric impulses. These impulses direct the activity of muscles and glands and, especially, perform higher intellectual functions.

ORGANS

Tissues are not distributed throughout the body at random. Instead, they are combined in very precise ways to create different structural units, the **organs**. These include the skin, the stomach, the liver, the lungs, the heart, and so on. These organs are responsible for carrying out specific tasks.

Each organ has a particular shape, a particular location, and a specific task. Some are solid, and others are hollow ducts. However, they all consist of various elements of tissue. Some organs contain tissue that is not present in any other part of the body, as is the case with the epidermis, the topmost layer of the skin, or osseous tissue, which is the main component of bones.

However, there are very different organs whose properties depend on the presence of the same kind of tissue—the numerous muscles of the body. The heart and various intestinal cavities, for example, can contract and relax because they consist of muscle tissue. What characterizes an organ, then, is not so much its anatomic composition but, rather, its function because each organ performs a specific activity that is indispensable to the whole.

APPARATUSES AND ORGANIC SYSTEMS

There are organs that perform certain specific functions on their own. For instance, the skin covers our whole bodies and offers protection to our internal structures, although it also has other roles. There are other organs that can perform their activities only in combination with other organs, to which they are intimately related. Together, these constitute a functional unit: an apparatus or system.

In reality, although the terms *apparatus* and *system* are used synonymously, there are certain subtle differences between them. An **apparatus** is when a series of organs is made up of different types of tissues. For example, the digestive apparatus is made up of elements as different as the mouth, the stomach, and the liver. The respiratory apparatus is made up of organs including the nose, the larynx, the bronchi, and the lungs. The circulatory apparatus consists of the heart, the arteries, and the veins. A **system**, in contrast, is when all the components consist of the same tissue. For example, the nervous system consists basically of nervous tissue. The osseous system and the muscular system consist of

The cell

The human body

The locomotive system

The digestive system

The respiratory system

The circulatory system

Blood

Lymph

The nervous system

The senses

The urinary system

The reproductive system

Human reproduction

The endocrine system

The immune system

Alphabetical index

osseous and muscle tissue, respectively, although both are part of the locomotive apparatus. The endocrine system consists of different glandular organs that secrete hormones into the blood.

All the apparatuses and systems, however, are **interrelated**. The functions of each one can be fully performed only in conjunction with the others. All of them are necessary to create an autonomous organism. To understand, use the above-mentioned apparatuses and systems as examples. The digestive system is responsible for nutrition. The respiratory system permits us to obtain oxygen from our environment. The circulatory system makes it possible for blood-carrying nutrients and oxygen to reach the tissues. The locomotive apparatus allows us to make the movements necessary for daily life. The nervous system, together with the endocrine system, regulates all our bodily activity. Equally important as these organs are many others, including the senses, the urinary apparatus, the reproductive system, and so on.

Following a succinct description of the components of cells and the different areas of the human body, this volume then covers each and every one of the apparatuses and systems. To simplify, the words *apparatus* and *system* will be used interchangeably throughout this text.

The difference between a **system** (*top*, diagram of the nervous system) and an **apparatus** (*bottom*, diagram of the circulatory apparatus) is that the components of a system are made up of the same type of tissue while the components of an apparatus are made up of different types of tissues.

THE CELL

The cell is the **smallest unit** in the human body. It is also the common feature of all forms of life. The most simple organisms, such as bacteria and protozoa, consist of a single cell, which lives independently. Our bodies, in contrast, are made up of **thousands of millions** of cells that function in coordination with one another. The cells of the different tissues and organs of the human body are found in very **different** shapes and sizes, but they all have a similar basic structure.

CELLULAR ORGANELLES

Organelle is the name given to the minute structures that float in the cytoplasm and carry out specific functions vital for cellular existence: synthesizing proteins, acquiring energy, digesting food, and so on. They are the cell's equivalent of the complex organs of the body.

COMPONENTS OF THE HUMAN CELL

microvilli
fine folds of cytoplasmic membrane that increase the surface area of the cell and are involved in the interchange of substances with the external environment

rough endoplasmic reticulum
system of membranes and microchannels containing numerous ribosomes

cellular membrane or cytoplasm
semipermeable cell covering through which exchanges take place between the interior of the cell and the external environment

vacuoles
small bags that are used for storing reserves or for expelling secretions

microfilaments
fine threads of protein linked to the internal currents of the cell and responsible for contracting muscular fibers

smooth endoplasmic reticulum
system of membranes and channels that aid the transport of substances within the cell

Golgi apparatus
a series of small sacs and tubules responsible for transforming, transporting, and eliminating the chemical products that are required for cellular activity; they are the cell's "production line"

centrioles
tubular organelles that are involved in the process of cell division

microtubules
tubular filaments that form a type of internal skeleton in the cell and help it maintain its shape

ribosome
grain-shaped organelle responsible for making proteins

lysosome
small sac containing enzymes, responsible for digesting food and breaking down cell residues

mitochondrion
organelle with an extended, partitioned form in which the combustion of nutrients takes place; this is the "power source" of the cell

nucleus
a spherical formation containing genetic material, which is responsible for the functioning of the cell and the transmission of hereditary characteristics

nucleolus
small, spherical body contained in the nucleus that sends messages to the ribosomes in the cytoplasm to produce more protein

cytoplasm
a gelatinous-type substance in which the nucleus and all the organelles are immersed

nuclear membrane
covering of the nucleus that separates it from the cytoplasm

Introduction

The cell

The human body

The locomotive system

The digestive system

The respiratory system

The circulatory system

Blood

Lymph

The nervous system

The senses

The urinary system

The reproductive system

Human reproduction

The endocrine system

The immune system

Alphabetical index

CELL NUCLEUS

nuclear membrane

chromatin

nucleolus

nuclear pore

Human cells are eukaryotic cells since they consist of a nucleus separated from the cytoplasm by a nuclear membrane. Inside the nucleus are the elements that contain hereditary information and that direct all the functions of the cell. This information is contained in molecules of deoxyribonucleic acid (DNA). This substance is found throughout the nucleus in the form of chromatin while the cell is nonactive but condenses and forms stick-shaped chromosomes when the cell divides.

CHROMOSOMES DURING CELL DIVISION

STRUCTURE OF A CHROMOSOME

long arm

centromere

short arm

DNA

Each chromosome basically consists of a filament of DNA of varying length. At its center is found a constriction, called the centromere, that divides the chromosome into two unequal lengths, one short, one long.

SCHEMATIC REPRESENTATION OF THE DNA CHAIN

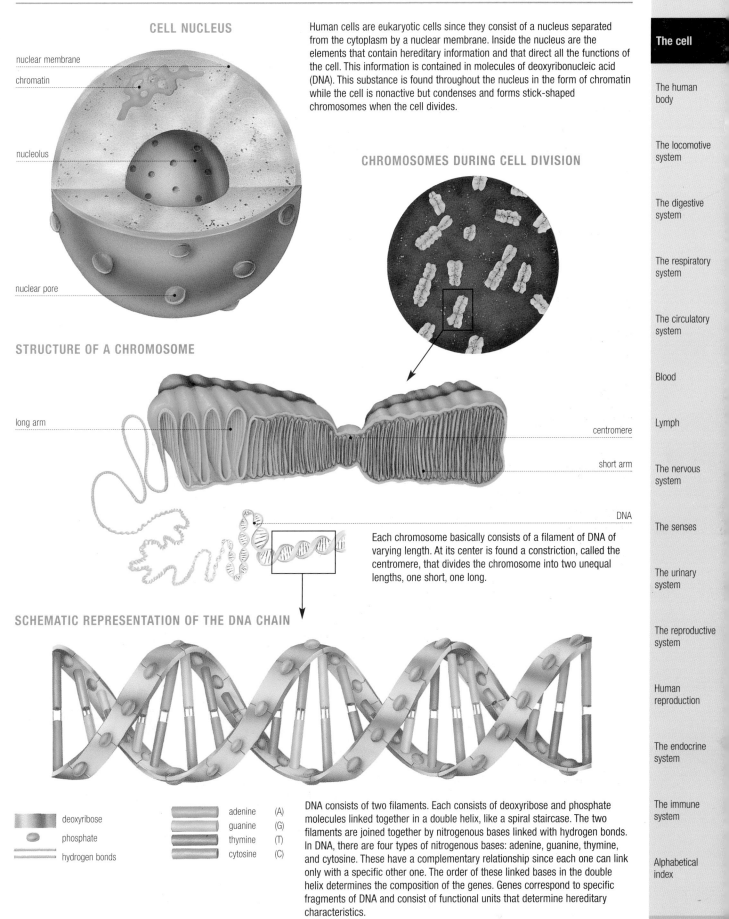

deoxyribose

phosphate

hydrogen bonds

adenine (A)
guanine (G)
thymine (T)
cytosine (C)

DNA consists of two filaments. Each consists of deoxyribose and phosphate molecules linked together in a double helix, like a spiral staircase. The two filaments are joined together by nitrogenous bases linked with hydrogen bonds. In DNA, there are four types of nitrogenous bases: adenine, guanine, thymine, and cytosine. These have a complementary relationship since each one can link only with a specific other one. The order of these linked bases in the double helix determines the composition of the genes. Genes correspond to specific fragments of DNA and consist of functional units that determine hereditary characteristics.

THE HUMAN BODY

Despite innumerable individual variations, the bodies of all human beings are very similar. They basically consist of the **head**, the **trunk**, and four **limbs**—two upper limbs (the arms) and two lower limbs (the legs). Of course, there are very clear **differences** between the male and the female body.

The male body is more muscular and sinewy, while the female body is more rounded and graceful. The greatest difference is in the genital organs and the secondary sexual characteristics, such as the distribution of body hair and the development of breasts.

MALE ANATOMY   ANTERIOR AND DORSAL VIEW

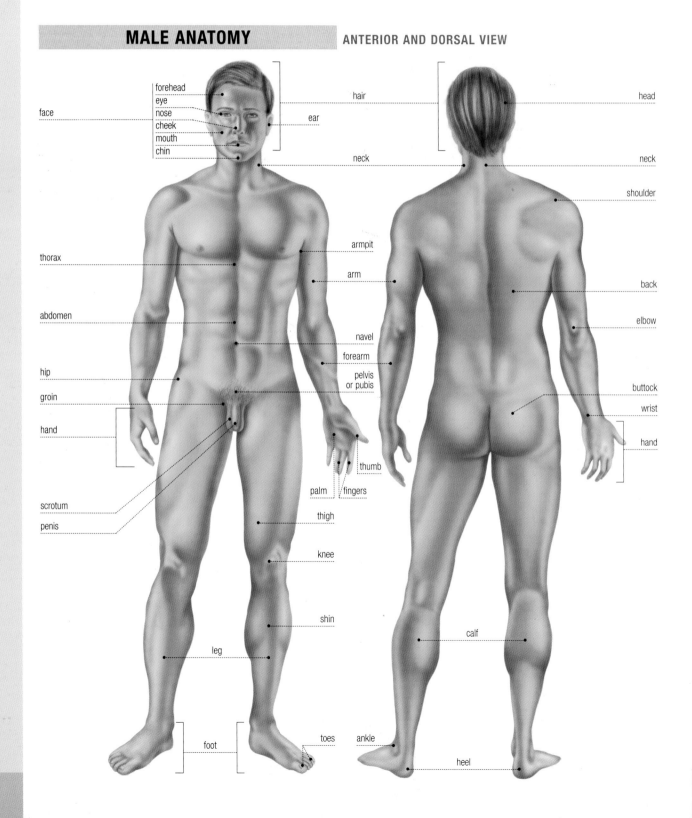

face

forehead
eye
nose
cheek
mouth
chin

ear

thorax

armpit

arm

abdomen

navel

hip

forearm

groin

pelvis or pubis

hand

scrotum

penis

thumb

palm   fingers

thigh

knee

shin

leg

foot   toes

hair

head

neck

neck

shoulder

back

elbow

buttock

wrist

hand

calf

ankle

heel

FEMALE ANATOMY

ANTERIOR AND DORSAL VIEW

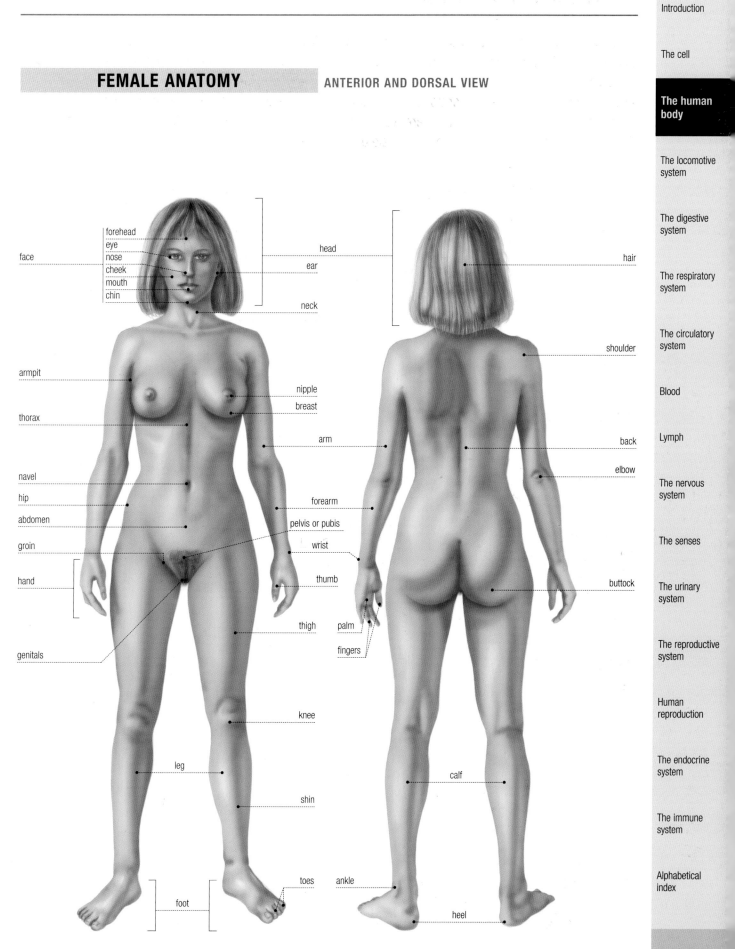

forehead
eye
nose
cheek
mouth
chin

face

head
ear

neck

armpit

thorax

nipple
breast

arm

navel

hip

abdomen

groin

hand

genitals

forearm

pelvis or pubis

wrist

thumb

thigh

palm

fingers

knee

leg

shin

toes

foot

ankle

heel

hair

shoulder

back

elbow

buttock

calf

Introduction

The cell

The human body

The locomotive system

The digestive system

The respiratory system

The circulatory system

Blood

Lymph

The nervous system

The senses

The urinary system

The reproductive system

Human reproduction

The endocrine system

The immune system

Alphabetical index

THE SKELETAL SYSTEM

Bones are strong, resistant elements of varying size and shape. They make up the **frame of the body** and make it possible for us to move. However, they are not inert elements. They are made of **living tissue** that is constantly active and on which minerals are deposited to give bone its particular consistency.

SCHEMATIC REPRESENTATION OF OSSEOUS TISSUE

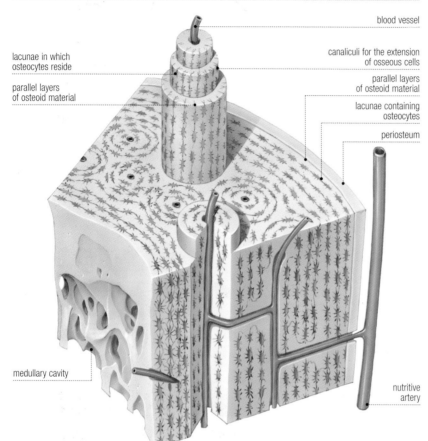

blood vessel

canaliculi for the extension of osseous cells

parallel layers of osteoid material

lacunae containing osteocytes

periosteum

lacunae in which osteocytes reside

parallel layers of osteoid material

medullary cavity

nutritive artery

Bone tissue is a complex framework of organic elements and minerals in a state of constant renovation. Specialized cells called **osteoblasts** manufacture an organic matrix of collagen fibers and a shapeless material, **osteoid material**, in which minerals such as calcium and phosphorous are deposited. When the osteoblasts are trapped in the osteoid material, they become **osteocytes**, which are inactive.

Osteoid material is arranged in concentric layers around a tube through which a blood vessel passes and through which run lots of small, transverse channels, or canaliculi. In this way, infinite numbers of small cross struts, called **trabeculae**, are formed. The number and quantity of these trabeculae differentiate two types of bone tissue. **Compact** bone is harder and forms the cortex of bones. **Cancellous** bone is less dense, is porous in appearance, and contains the bone medulla.

At birth, bones are made of **cartilage**, which is gradually replaced by bone tissue.

COMPONENTS OF OSSEOUS TISSUE

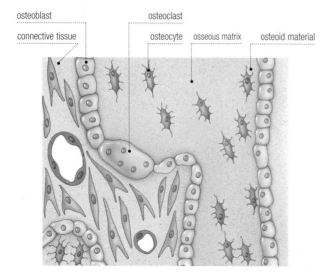

osteoblast

osteoclast

connective tissue

osteocyte   osseous matrix   osteoid material

FORMATION AND GROWTH OF BONE

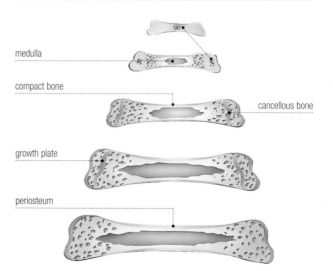

medulla

compact bone

cancellous bone

growth plate

periosteum

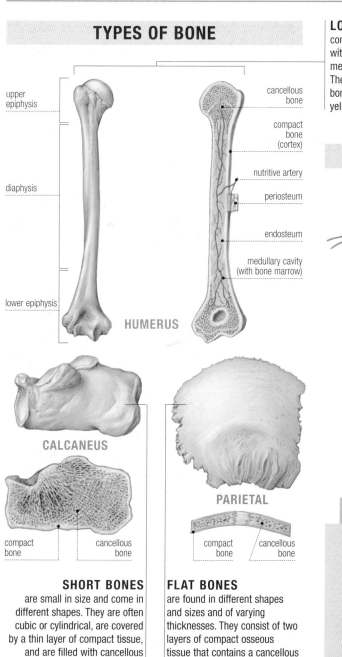

TYPES OF BONE

upper epiphysis

diaphysis

lower epiphysis

HUMERUS

cancellous bone

compact bone (cortex)

nutritive artery

periosteum

endosteum

medullary cavity (with bone marrow)

CALCANEUS

compact bone — cancellous bone

PARIETAL

compact bone | cancellous bone

SHORT BONES
are small in size and come in different shapes. They are often cubic or cylindrical, are covered by a thin layer of compact tissue, and are filled with cancellous bone tissue.

FLAT BONES
are found in different shapes and sizes and of varying thicknesses. They consist of two layers of compact osseous tissue that contains a cancellous osseous material called diploe.

LONG BONES
consist of a central body (diaphysis) and two extremes (epiphysis), with an external layer of compact tissue (cortex) covered in a hard membrane (periosteum) and a strong internal membrane (endosteum). The extremities consist of cancellous bone tissue, which contains red bone marrow (medulla). In the main body, there is a cavity that contains yellow bone marrow.

OSSEOUS BLOOD SUPPLY

Irrigation of the surface of the bone is performed by periosteal arteries. The interior of the bone is supplied by nutritive arteries, which penetrate the bone and divide into infinite small branches.

periosteal artery

nutritive artery

BONE FUNCTION

• Bones form the rigid structure that provides the organism with a frame, determining its size and form.

• Bones protect various soft internal organs that are vulnerable to blows and external attack.

• Bones are the rigid elements of the locomotive system. They provide support points for muscles and thus provide points of leverage that allow different parts of the body to be moved.

• Bones represent an important reserve of minerals such as calcium and phosphorous.

• Bones contain the bone marrow, or medulla, where blood cells are produced.

BONE FRACTURES

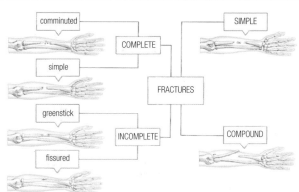

comminuted

COMPLETE

simple

SIMPLE

greenstick

FRACTURES

fissured

INCOMPLETE

COMPOUND

A bone fracture is when a **bone is broken**. It is either **incomplete**, when the fracture is partial, or **complete**, when the bone is broken into two or more pieces and sometimes into several pieces (a comminuted fracture). If the skin covering the bone is undamaged, the fracture is **simple**. In contrast, if the surface tissue tears and fragments of the broken bone are open to the exterior, the fracture is **compound**.

Introduction

The cell

The human body

The locomotive system

The digestive system

The respiratory system

The circulatory system

Blood

Lymph

The nervous system

The senses

The urinary system

The reproductive system

Human reproduction

The endocrine system

The immune system

Alphabetical index

THE SKELETAL SYSTEM

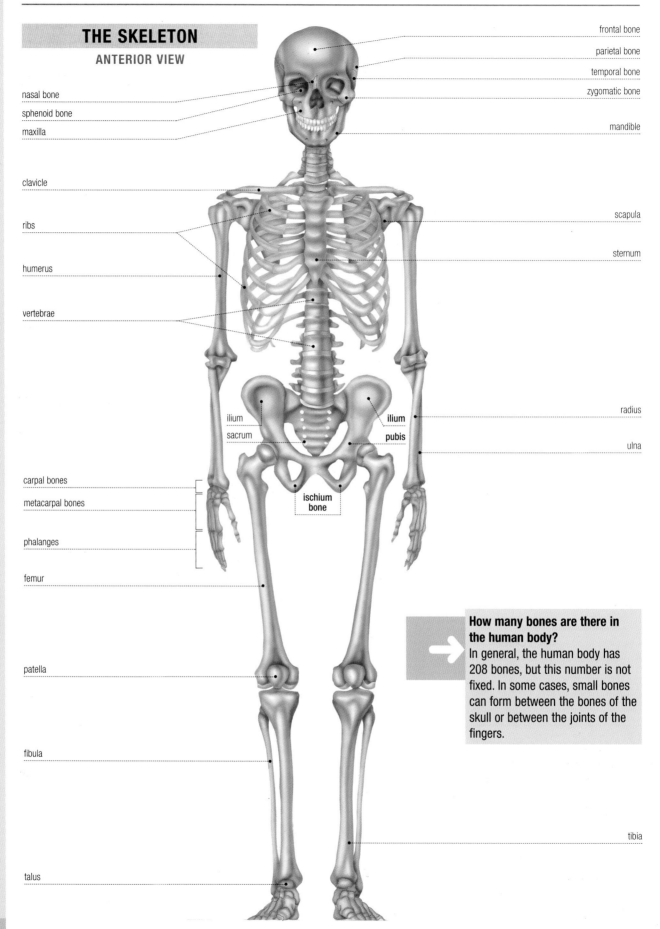

THE SKELETON
ANTERIOR VIEW

nasal bone
sphenoid bone
maxilla

clavicle

ribs

humerus

vertebrae

carpal bones

metacarpal bones

phalanges

femur

patella

fibula

talus

frontal bone
parietal bone
temporal bone
zygomatic bone

mandible

scapula

sternum

ilium

ilium

sacrum

pubis

radius

ulna

ischium
bone

tibia

How many bones are there in the human body?
In general, the human body has 208 bones, but this number is not fixed. In some cases, small bones can form between the bones of the skull or between the joints of the fingers.

Introduction

The cell

The human body

The locomotive system

The digestive system

The respiratory system

The circulatory system

Blood

Lymph

The nervous system

The senses

The urinary system

The reproductive system

Human reproduction

The endocrine system

The immune system

Alphabetical index

POSTERIOR VIEW

parietal bone

temporal bone

clavicle

scapula

ribs

sacrum

coccyx

femur

occipital bone

mandible

vertebrae

ilium

ulna

radius

carpal bones

metacarpal bones

phalanges

tibia

tibula

talus

calcaneus

Precise symmetry

The skeleton is exactly symmetrical. If you divide it lengthways along a vertical line down the middle, both sides are identical. For this reason, all the bones that are unique are symmetrical, while all the bones that come in pairs are irregular but are located symmetrically opposite one another, on different sides of the body.

THE SKELETAL SYSTEM

THE SKULL BONES
ANTERIOR VIEW

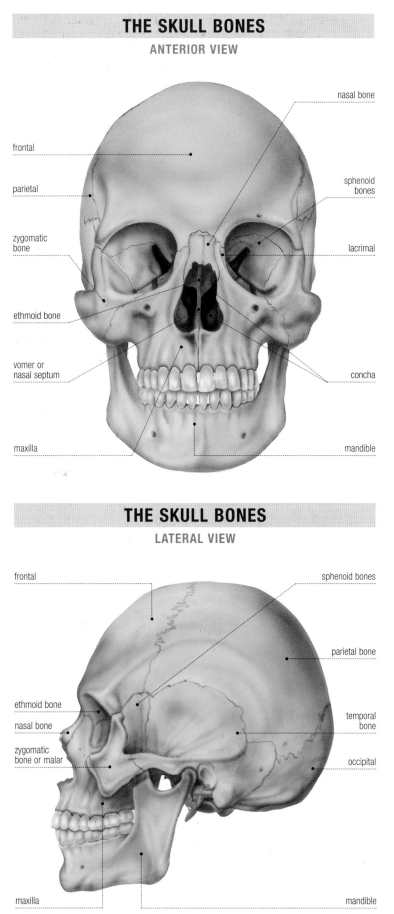

nasal bone

frontal

parietal

sphenoid bones

zygomatic bone

lacrimal

ethmoid bone

vomer or nasal septum

concha

maxilla

mandible

THE BONES OF THE HEAD ARE DIVIDED INTO TWO PARTS:

• The **skull** consists of the posterior part and is made up of eight bones fused together to create a hollow that contains the brain.

• The **face** consists of the anterior part and is made up of different bones joined together, except for the mandible, which moves. It holds most of the sense organs and contains the beginnings of the respiratory and digestive systems.

THE SMALL BONES OF THE MIDDLE EAR

In the middle ear, which is contained within the thickness of the temporal bone, are three small bones in a chain. They do not actually form part of the skeleton, although they play a fundamental part in hearing. They are the malleus, the incus, and the stapes.

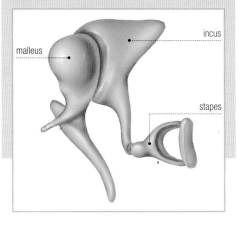

incus

malleus

stapes

THE SKULL BONES
LATERAL VIEW

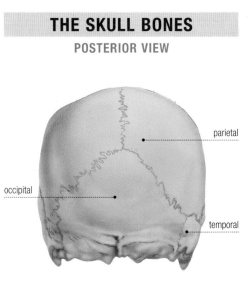

frontal

sphenoid bones

ethmoid bone

parietal bone

nasal bone

zygomatic bone or malar

temporal bone

occipital

maxilla

mandible

THE SKULL BONES
POSTERIOR VIEW

parietal

occipital

temporal

VERTEBRAL COLUMN

LATERAL VIEW

The vertebral column constitutes the **truncal axis**. It stretches along the length of the midline of the back, from the base of the skull to the pelvis. It consists of a series of bones, the **vertebrae**, placed one on top of the other. In total, there are 34 vertebrae, but only the 24 upper ones are independent, while the bottom ones are fused together and form the bones of the **sacrum** and the **coccyx**.

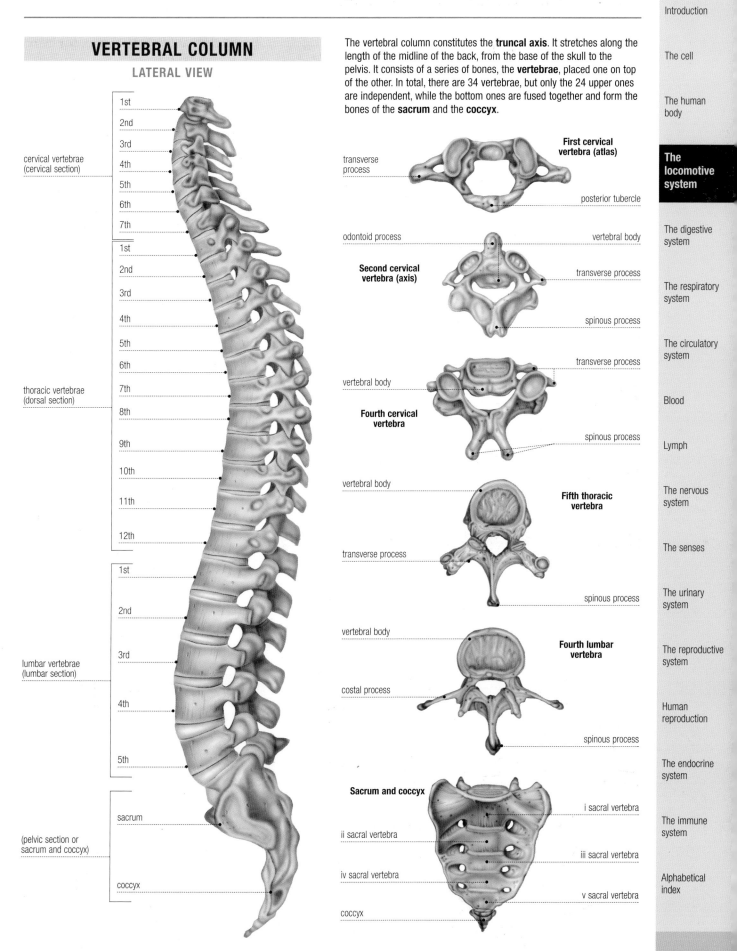

cervical vertebrae (cervical section)
- 1st
- 2nd
- 3rd
- 4th
- 5th
- 6th
- 7th

thoracic vertebrae (dorsal section)
- 1st
- 2nd
- 3rd
- 4th
- 5th
- 6th
- 7th
- 8th
- 9th
- 10th
- 11th
- 12th

lumbar vertebrae (lumbar section)
- 1st
- 2nd
- 3rd
- 4th
- 5th

(pelvic section or sacrum and coccyx)
- sacrum
- coccyx

First cervical vertebra (atlas)
- transverse process
- posterior tubercle

Second cervical vertebra (axis)
- odontoid process
- vertebral body
- transverse process
- spinous process

Fourth cervical vertebra
- vertebral body
- transverse process
- spinous process

Fifth thoracic vertebra
- vertebral body
- transverse process
- spinous process

Fourth lumbar vertebra
- vertebral body
- costal process
- spinous process

Sacrum and coccyx
- ii sacral vertebra
- iv sacral vertebra
- coccyx
- i sacral vertebra
- iii sacral vertebra
- v sacral vertebra

Introduction

The cell

The human body

The locomotive system

The digestive system

The respiratory system

The circulatory system

Blood

Lymph

The nervous system

The senses

The urinary system

The reproductive system

Human reproduction

The endocrine system

The immune system

Alphabetical index

THE SKELETAL SYSTEM

RIGHT SCAPULA
ANTERIOR VIEW

The skeleton of the upper limb consists of the bones of the **arm** (humerus), **forearm** (ulna and radius), and **hand** (carpus, metacarpus, and fingers).

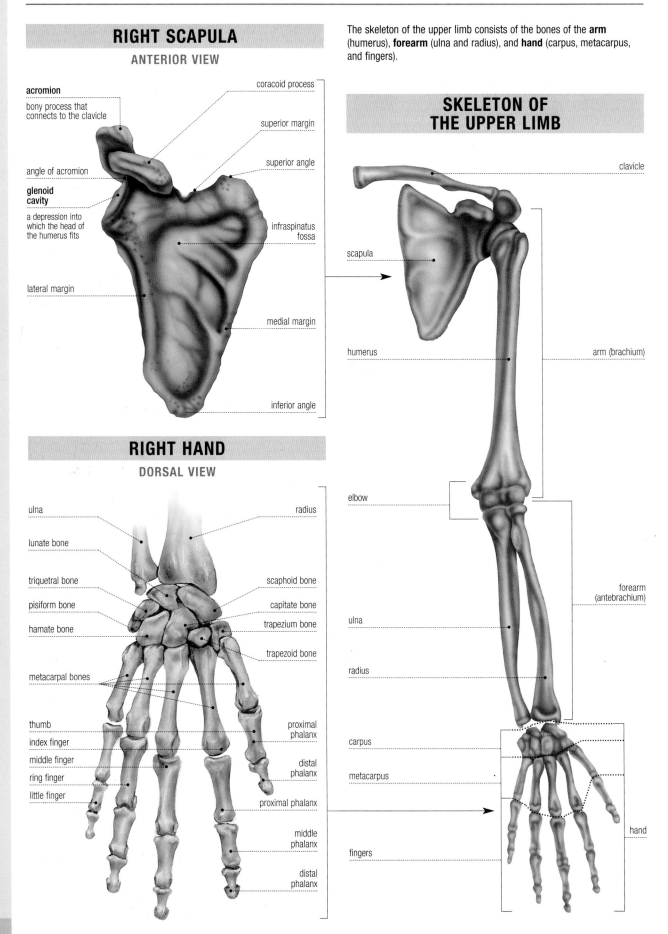

acromion
bony process that connects to the clavicle

coracoid process

superior margin

superior angle

angle of acromion

glenoid cavity
a depression into which the head of the humerus fits

infraspinatus fossa

lateral margin

medial margin

inferior angle

SKELETON OF THE UPPER LIMB

clavicle

scapula

humerus

arm (brachium)

elbow

forearm (antebrachium)

ulna

radius

carpus

metacarpus

fingers

hand

RIGHT HAND
DORSAL VIEW

ulna

radius

lunate bone

triquetral bone

scaphoid bone

pisiform bone

capitate bone

hamate bone

trapezium bone

trapezoid bone

metacarpal bones

thumb

index finger

proximal phalanx

middle finger

ring finger

distal phalanx

little finger

proximal phalanx

middle phalanx

distal phalanx

The skeleton of the lower limb consists of the bones of the **thigh** (femur), **leg** (tibia and fibula), and **foot** (tarsus, metatarsus, and toes).

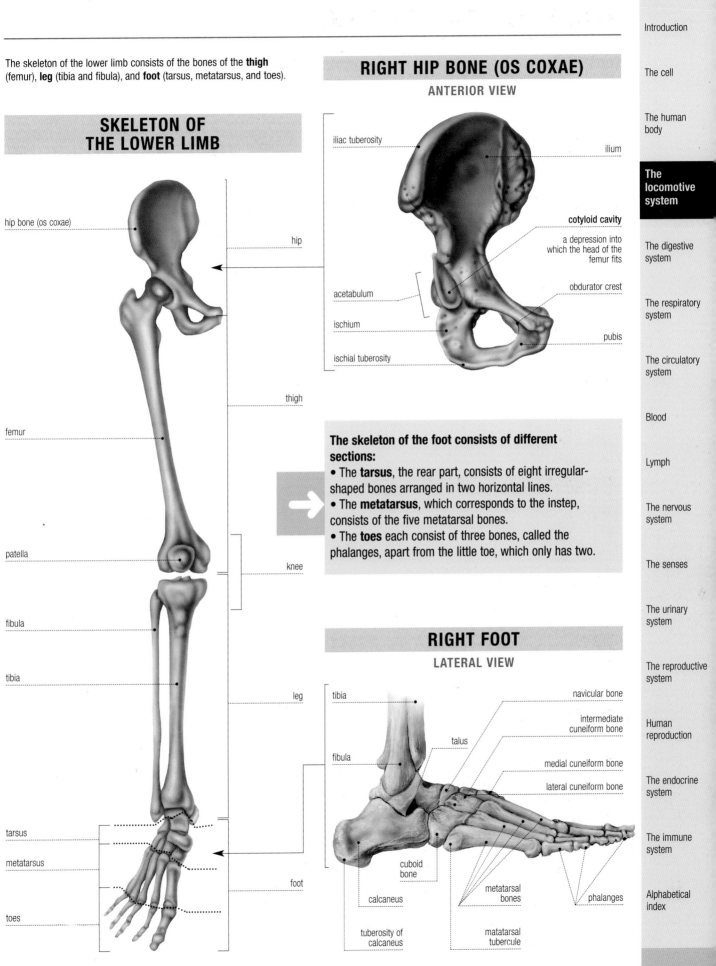

SKELETON OF THE LOWER LIMB

hip bone (os coxae)

hip

thigh

femur

patella

knee

fibula

tibia

leg

tarsus

metatarsus

foot

toes

RIGHT HIP BONE (OS COXAE)
ANTERIOR VIEW

iliac tuberosity

ilium

cotyloid cavity

a depression into which the head of the femur fits

acetabulum

obdurator crest

ischium

pubis

ischial tuberosity

The skeleton of the foot consists of different sections:
• The **tarsus**, the rear part, consists of eight irregular-shaped bones arranged in two horizontal lines.
• The **metatarsus**, which corresponds to the instep, consists of the five metatarsal bones.
• The **toes** each consist of three bones, called the phalanges, apart from the little toe, which only has two.

RIGHT FOOT
LATERAL VIEW

tibia

navicular bone

intermediate cuneiform bone

talus

fibula

medial cuneiform bone

lateral cuneiform bone

cuboid bone

metatarsal bones

phalanges

calcaneus

tuberosity of calcaneus

matatarsal tubercule

Introduction

The cell

The human body

The locomotive system

The digestive system

The respiratory system

The circulatory system

Blood

Lymph

The nervous system

The senses

The urinary system

The reproductive system

Human reproduction

The endocrine system

The immune system

Alphabetical index

JOINTS

Joints are the **points of contact** between the different bones that make up the skeleton. The human body consists of more than 200 joints, but they are of different types and have different functions. Some are responsible for the movement of the various parts of the skeleton. Others, on the other hand, are not very mobile or are fixed. Their purpose is to support and hold together different parts of the skeleton.

TYPES OF JOINTS

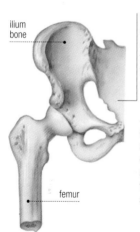

synarthrosis
these are fixed joints, incapable of movement; they consist of a solid union between two or more pieces of bone and, in this way, form a protective layer for the soft tissue that they cover

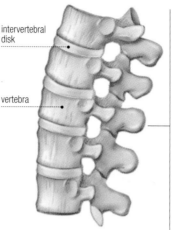

intervertebral disk

vertebra

symphysis
these joints have a small degree of movement; the bones are not directly joined to one another; they are separated by fibrocartilage whose consistency allows it to change shape temporarily to afford a certain degree of movement of the bony sections

diarthrosis
these are different types of joints that allow a wide range of movements

ilium bone

femur

enarthrosis
a moving joint that consists of a spherical section of bone that fits within a cavity and therefore can move in all directions

condylarthrosis
a moving joint that consists of one rounded or elliptical section of bone and another with a hollow that acts as a socket for the first

humerus

radius

fibula

arthodial
a moving joint that consists of sections of flat bones that can slide only over one another

atlas

axis

trochlearthrosis
a moving joint that consists of a section of bone in the shape of a pulley with a depression in the center and a bone with a ridge or process that fits into the channel of the pulley

DISLOCATIONS

Luxations, or dislocations, consist of a **displacement** of the sections of bone that form a joint so that the parts of the joint are no longer properly connected together.

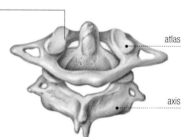

clavicle
scapula
joint cavity
humerus
head of disjointed humerus

LUXATION OF THE SHOULDER

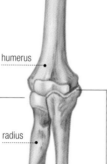

humerus
radius
ulna
disjointed humerus

LUXATION OF THE ELBOW

first phalanx
second phalanx
third phalanx

INTERPHALANGAL LUXATION OF THE FINGER

hip bone
joint cavity
femur
head of disjointed femur

LUXATION OF THE HIP

ELEMENTS OF A MOBILE JOINT

In a mobile joint, in addition to the bones that are joined together, there are various elements designed to protect the ends of the bones as well as other elements that ensure that the joint remains stable:

• **articular cartilage:** a thin layer of strong, elastic tissue that covers the ends of the bones and prevents them from rubbing together so that they do not wear down;

• **articular capsule:** a fibrous membrane that covers the whole joint and prevents the different sections of bone from moving too far;

• **synovial membrane:** a layer of smooth, shiny tissue that covers the inside of an articular capsule and secretes a viscous liquid that fills the joint, ensuring it is lubricated and providing nutrition for the articular cartilage;

• **ligaments:** strong, fibrous bands that provide stability for the joint.

KNEE JOINT
LATERAL SECTION

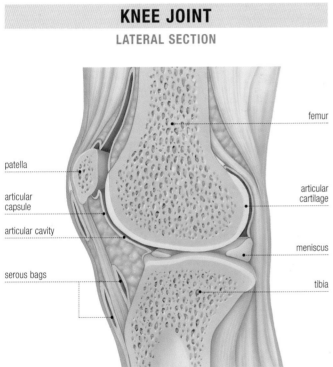

femur

patella

articular capsule

articular cavity

articular cartilage

serous bags

meniscus

tibia

SHOULDER JOINT
ANTERIOR SECTION

articular capsule

ligament

humerus

synovial membrane

scapula

articular cartilage

articular liquid

articular liquid

synovial membrane

articular capsule

HIP BONE
ANTERIOR VIEW

hip bone

iliofemoral ligament

articular capsule

pubofemoral ligament

femur

SECTION

hip bone

cotyloid flange

articular capsule

cotyloid cavity

round ligament

femur

MENISCUS

tibia

internal meniscus

external meniscus

A meniscus consists of **fibrous cartilage** that is found between the bony sections of some joints. It increases the area of contact between the bones, distributes pressure better, and limits extreme movements. There are menisci in various joints, but the most important ones are found in the knee.

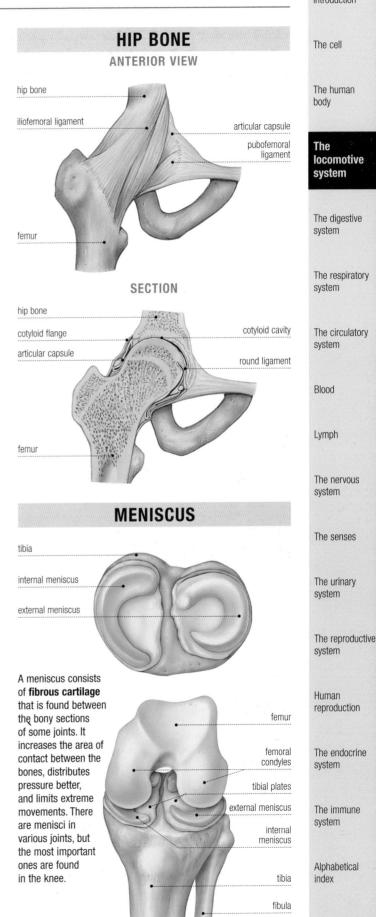

femur

femoral condyles

tibial plates

external meniscus

internal meniscus

tibia

fibula

Introduction

The cell

The human body

The locomotive system

The digestive system

The respiratory system

The circulatory system

Blood

Lymph

The nervous system

The senses

The urinary system

The reproductive system

Human reproduction

The endocrine system

The immune system

Alphabetical index

THE MUSCULAR SYSTEM

Muscles are very special **masses of tissue** because they are able to contract and relax, which changes their length. There are different types of muscles. The ones called skeletal muscles, which are connected to bones either directly or by means of fibrous bands (tendons), can be contracted at will so that we can move different parts of our bodies. Thanks to the action of our muscles, we can walk and jump, hold on to objects or let them go, hit or caress, chew and whistle, scratch our noses, and so on.

MUSCLE STRUCTURE

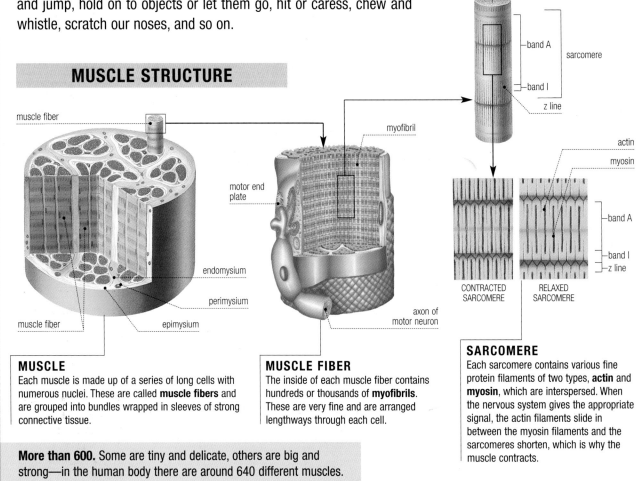

MYOFIBRIL
When looked at under an electron microscope, a series of regular grooves can be seen. These form bands of different tones and make up the functional units of the muscle, the **sarcomeres**.

MUSCLE
Each muscle is made up of a series of long cells with numerous nuclei. These are called **muscle fibers** and are grouped into bundles wrapped in sleeves of strong connective tissue.

MUSCLE FIBER
The inside of each muscle fiber contains hundreds or thousands of **myofibrils**. These are very fine and are arranged lengthways through each cell.

SARCOMERE
Each sarcomere contains various fine protein filaments of two types, **actin** and **myosin**, which are interspersed. When the nervous system gives the appropriate signal, the actin filaments slide in between the myosin filaments and the sarcomeres shorten, which is why the muscle contracts.

More than 600. Some are tiny and delicate, others are big and strong—in the human body there are around 640 different muscles.

MUSCLE SHAPES

Although all muscles are made up of the same components and act in a similar way, they are very diverse in shape, and each is adapted to its specific function.

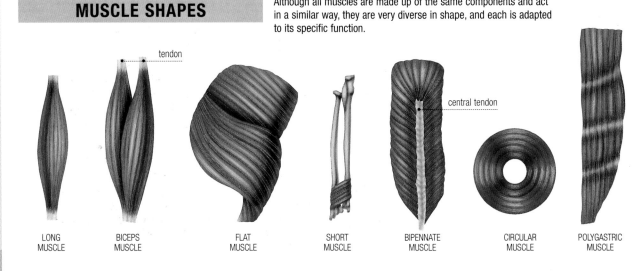

| LONG MUSCLE | BICEPS MUSCLE | FLAT MUSCLE | SHORT MUSCLE | BIPENNATE MUSCLE | CIRCULAR MUSCLE | POLYGASTRIC MUSCLE |

MUSCLES OF THE HUMAN BODY ANTERIOR VIEW

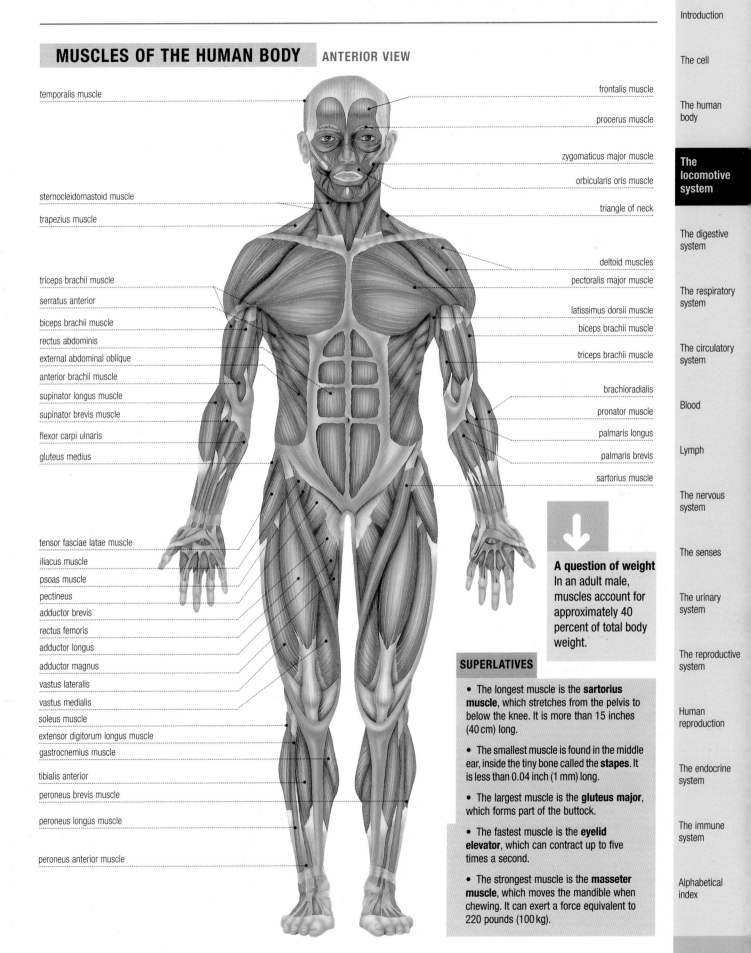

temporalis muscle

frontalis muscle

procerus muscle

zygomaticus major muscle

orbicularis oris muscle

triangle of neck

sternocleidomastoid muscle

trapezius muscle

deltoid muscles

pectoralis major muscle

triceps brachii muscle

serratus anterior

latissimus dorsii muscle

biceps brachii muscle

biceps brachii muscle

rectus abdominis

triceps brachii muscle

external abdominal oblique

anterior brachii muscle

brachioradialis

supinator longus muscle

pronator muscle

supinator brevis muscle

palmaris longus

flexor carpi ulnaris

palmaris brevis

gluteus medius

sartorius muscle

tensor fasciae latae muscle

iliacus muscle

psoas muscle

pectineus

adductor brevis

rectus femoris

adductor longus

adductor magnus

vastus lateralis

vastus medialis

soleus muscle

extensor digitorum longus muscle

gastrocnemius muscle

tibialis anterior

peroneus brevis muscle

peroneus longus muscle

peroneus anterior muscle

A question of weight

In an adult male, muscles account for approximately 40 percent of total body weight.

SUPERLATIVES

• The longest muscle is the **sartorius muscle**, which stretches from the pelvis to below the knee. It is more than 15 inches (40 cm) long.

• The smallest muscle is found in the middle ear, inside the tiny bone called the **stapes**. It is less than 0.04 inch (1 mm) long.

• The largest muscle is the **gluteus major**, which forms part of the buttock.

• The fastest muscle is the **eyelid elevator**, which can contract up to five times a second.

• The strongest muscle is the **masseter muscle**, which moves the mandible when chewing. It can exert a force equivalent to 220 pounds (100 kg).

Introduction

The cell

The human body

The locomotive system

The digestive system

The respiratory system

The circulatory system

Blood

Lymph

The nervous system

The senses

The urinary system

The reproductive system

Human reproduction

The endocrine system

The immune system

Alphabetical index

THE MUSCULAR SYSTEM

MUSCLES OF THE HUMAN BODY
POSTERIOR VIEW

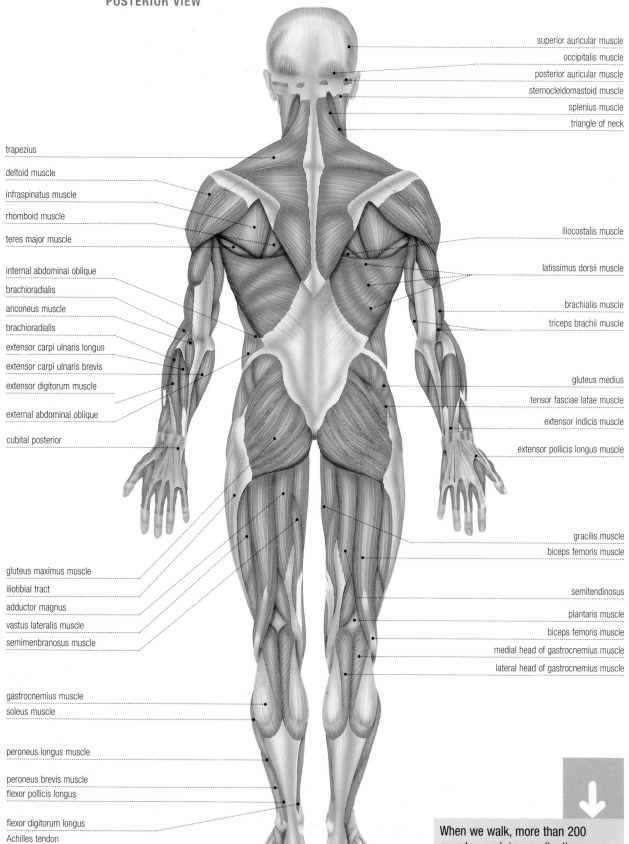

superior auricular muscle

occipitalis muscle

posterior auricular muscle

sternocleidomastoid muscle

splenius muscle

triangle of neck

trapezius

deltoid muscle

infraspinatus muscle

rhomboid muscle

teres major muscle

iliocostalis muscle

latissimus dorsii muscle

internal abdominal oblique

brachioradialis

anconeus muscle

brachioradialis

extensor carpi ulnaris longus

extensor carpi ulnaris brevis

extensor digitorum muscle

external abdominal oblique

cubital posterior

brachialis muscle

triceps brachii muscle

gluteus medius

tensor fasciae latae muscle

extensor indicis muscle

extensor pollicis longus muscle

gracilis muscle

biceps femoris muscle

gluteus maximus muscle

iliotibial tract

adductor magnus

vastus lateralis muscle

semimenbranosus muscle

semitendinosus

plantaris muscle

biceps femoris muscle

medial head of gastrocnemius muscle

lateral head of gastrocnemius muscle

gastrocnemius muscle

soleus muscle

peroneus longus muscle

peroneus brevis muscle

flexor pollicis longus

flexor digitorum longus

Achilles tendon

When we walk, more than 200 muscles work in coordination.

There are many muscles in the head. Some cover the skull and have limited movement, while others on the face are very mobile. These can be divided into two groups:

- **facial muscles:** these allow us to adopt different expressions and to express our emotions.

- **masticatory muscles:** these are responsible for movement of the maxilla, or jawbone.

MUSCLES OF THE HUMAN HEAD
ANTERIOR VIEW

frontalis muscle

temporalis muscle

orbicularis oculi muscle

pyramidalis nasi

transverse part of nasalis muscle

levator labii superioris alaeque nasi muscle

zygomaticus minor muscle

levator labii superioris muscle

masseter

depressor septi muscle

levator anguli oris muscle

zygomaticus major muscle

orbicularis oris muscle

buccinator

risorius muscle

depressor anguli oris muscle

platysma

depressor labii inferioris muscle

mentalis muscle

MUSCLES OF THE HUMAN HEAD
LATERAL VIEW

frontalis muscle

orbicularis oculi muscle

pyramidalis nasi

temporalis muscle

levator labii superioris alaeque nasi muscle

occipitalis muscle

transverse part of nasalis muscle

posterior auricular muscle

levator labii superioris muscle

sternocleidomastoid muscle

zygomaticus minor muscle

semispinalis capitis muscle

depressor septi muscle

trapezius muscle

levator anguli oris muscle

masseter

orbicularis oris muscle

splenius capitis muscle

risorius muscle

depressor labii inferioris muscle

zygomaticus major muscle

platysma

mentalis muscle

depressor anguli oris muscle

buccinator

Introduction

The cell

The human body

The locomotive system

The digestive system

The respiratory system

The circulatory system

Blood

Lymph

The nervous system

The senses

The urinary system

The reproductive system

Human reproduction

The endocrine system

The immune system

Alphabetical index

27

THE MUSCULAR SYSTEM

The upper limbs are moved by means of **large**, **strong** muscles such as the deltoids, which allow us to move our arms in all directions, or the biceps and triceps, which are responsible for **bending** and **stretching** the forearm. We also have smaller, thinner muscles that allow us to make small, precise movements using our fingers.

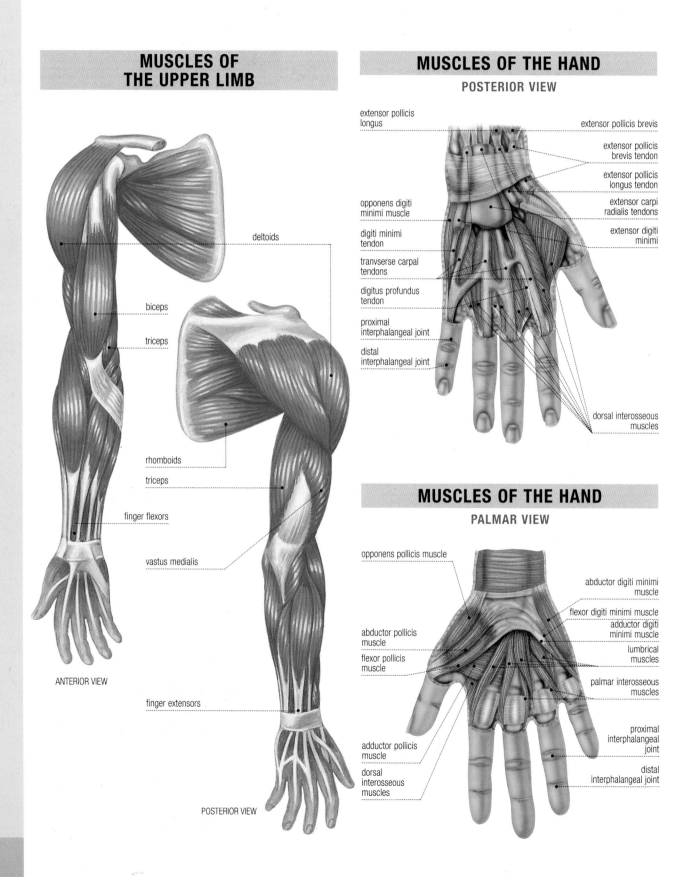

MUSCLES OF THE UPPER LIMB

deltoids

biceps

triceps

rhomboids

triceps

finger flexors

vastus medialis

ANTERIOR VIEW

finger extensors

POSTERIOR VIEW

MUSCLES OF THE HAND
POSTERIOR VIEW

extensor pollicis longus

opponens digiti minimi muscle

digiti minimi tendon

tranvserse carpal tendons

digitus profundus tendon

proximal interphalangeal joint

distal interphalangeal joint

extensor pollicis brevis

extensor pollicis brevis tendon

extensor pollicis longus tendon

extensor carpi radialis tendons

extensor digiti minimi

dorsal interosseous muscles

MUSCLES OF THE HAND
PALMAR VIEW

opponens pollicis muscle

abductor pollicis muscle

flexor pollicis muscle

adductor pollicis muscle

dorsal interosseous muscles

abductor digiti minimi muscle

flexor digiti minimi muscle

adductor digiti minimi muscle

lumbrical muscles

palmar interosseous muscles

proximal interphalangeal joint

distal interphalangeal joint

The muscles of the lower limbs are essential for **walking** and to keep us in an **upright position** on our feet. The largest muscles are the gluteus muscles, which make up the fleshy mass of the buttocks, and the muscles that make up the quadriceps femoris (rectus femoris, vastus lateralis, vastus medialis, and vastus intermedius).

MUSCLES OF THE LOWER LIMB

gluteus maximus

sartorius

adductor maximus

rectus femoris

biceps femoris

vastus medialis

vastus lateralis

gastrocnemius

soleus

Achilles tendon

POSTERIOR VIEW

ANTERIOR VIEW

MUSCLES OF THE FOOT
DORSAL VIEW

extensor digitorum longus tendon

extensor hallucis longus muscle

extensor digitorum longus tendons

tibialis anterior tendon

extensor digitorum brevis muscles

abductor peroneus longus muscle

extensor hallucis longus tendon

adductor hallucis muscle

dorsal interosseous muscles

MUSCLES OF THE FOOT
LATERAL VIEW

Achilles tendon

extensor hallucis longus muscle

tibialis anterior muscle

extensor digitorum longus tendon (section)

extensor hallucis longus tendon

lateral peroneus longus tendon

dorsal interosseous muscles

lateral peroneus brevis tendon

extensor digitorum brevis muscles

peroneus tendon

Introduction

The cell

The human body

The locomotive system

The digestive system

The respiratory system

The circulatory system

Blood

Lymph

The nervous system

The senses

The urinary system

The reproductive system

Human reproduction

The endocrine system

The immune system

Alphabetical index

THE DIGESTIVE SYSTEM

The digestive system has an extremely important function. It is responsible for **transforming the food** we eat each day so that the body can obtain the energy and nutritional elements it needs to create and maintain its tissues and ensure that its metabolism and various vital functions can operate correctly.

THE DIGESTIVE PROCESS

buccal cavity

tongue

pharynx

mouth

1 The digestive process starts with the mouth, where food is broken up by the action of the teeth and then submitted to the action of saliva.

teeth

esophagus

2 The bolus of food passes through the esophagus, by means of swallowing, to arrive at the stomach.

stomach

3 Food is temporarily stored in the stomach, where it is submitted to the powerful effects of gastric juice, which is produced in the glands of the mucous membrane that lines the stomach.

liver

4 The liver secretes bile, which is vital for fat absorption; stores vitamins, proteins, and glycogen; is involved in the metabolism of lipids; synthesizes proteins; and transforms toxic substances into harmless compounds.

pancreas

6 Pancreatic juice is involved in digesting all organic nutrients.

gallbladder

5 The gallbladder stores bile salts and later empties them into the duodenum.

transverse colon

duodenum

7 The duodenum is where pancreatic, bile, and intestinal juices collect, breaking down food and absorbing nutrients.

small intestine

8 The digestion and absorption of nutrients takes place along the whole length of the small intestine.

descending colon

jejunum

ascending colon

ileum

sigmoid colon

9 The bolus of food ends its long journey in the colon, where it is transformed into fecal matter.

rectum

10 Undigested detritus and residues from the digestive process are expelled through the rectum and anus.

vermiform appendix

anal duct

anus

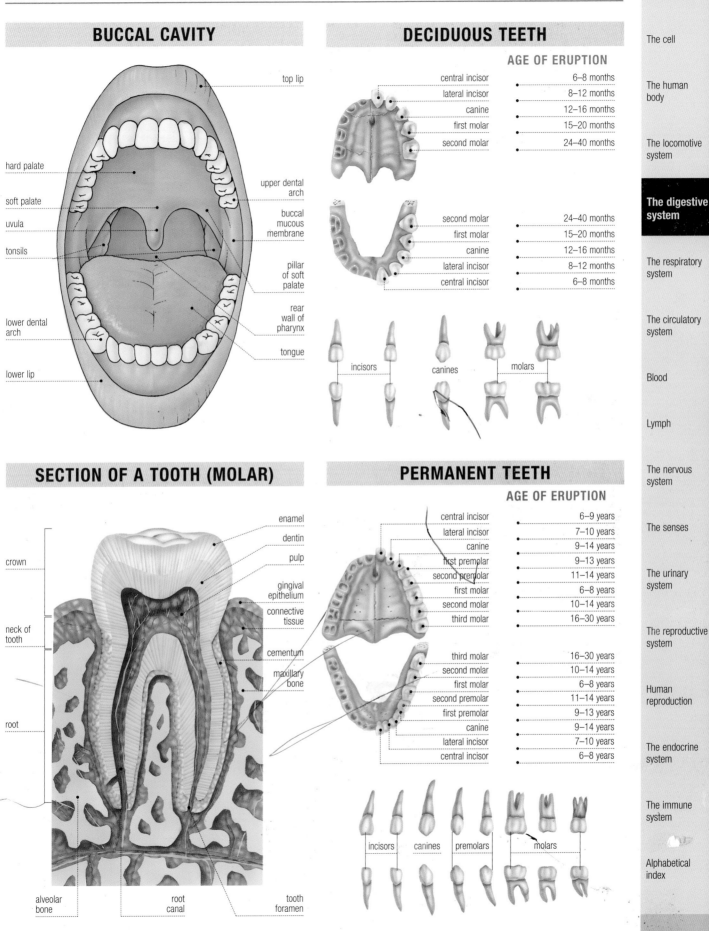

BUCCAL CAVITY

top lip

hard palate

soft palate

uvula

tonsils

upper dental arch

buccal mucous membrane

pillar of soft palate

rear wall of pharynx

tongue

lower dental arch

lower lip

DECIDUOUS TEETH

AGE OF ERUPTION

central incisor	6–8 months
lateral incisor	8–12 months
canine	12–16 months
first molar	15–20 months
second molar	24–40 months

second molar	24–40 months
first molar	15–20 months
canine	12–16 months
lateral incisor	8–12 months
central incisor	6–8 months

incisors   canines   molars

SECTION OF A TOOTH (MOLAR)

enamel

dentin

pulp

gingival epithelium

connective tissue

cementum

maxillary bone

crown

neck of tooth

root

alveolar bone

root canal

tooth foramen

PERMANENT TEETH

AGE OF ERUPTION

central incisor	6–9 years
lateral incisor	7–10 years
canine	9–14 years
first premolar	9–13 years
second premolar	11–14 years
first molar	6–8 years
second molar	10–14 years
third molar	16–30 years

third molar	16–30 years
second molar	10–14 years
first molar	6–8 years
second premolar	11–14 years
first premolar	9–13 years
canine	9–14 years
lateral incisor	7–10 years
central incisor	6–8 years

incisors   canines   premolars   molars

Introduction

The cell

The human body

The locomotive system

The digestive system

The respiratory system

The circulatory system

Blood

Lymph

The nervous system

The senses

The urinary system

The reproductive system

Human reproduction

The endocrine system

The immune system

Alphabetical index

THE ESOPHAGUS

ANTERIOR VIEW

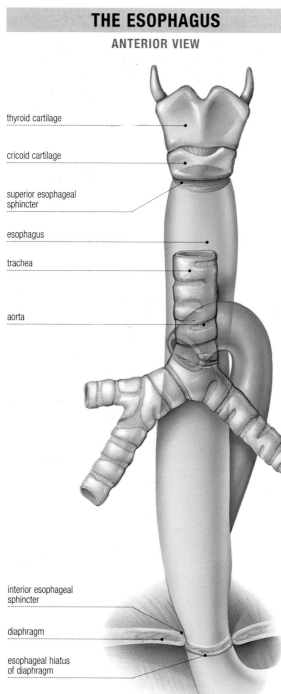

thyroid cartilage

cricoid cartilage

superior esophageal sphincter

esophagus

trachea

aorta

interior esophageal sphincter

diaphragm

esophageal hiatus of diaphragm

stomach

The esophagus is a tube about 10 inches (25 cm) long with muscular walls whose function is to transport food from the throat to the stomach. It begins at the pharynx, passes through the thoracic cavity from top to bottom, passes through the diaphragm into the abdominal cavity, and enters the gastric chamber.

SWALLOWING

The act of swallowing is a complex process. The first part is conscious and voluntary. After chewing the food, the tongue presses the bolus of food toward the palate and pushes it toward the pharynx (1). Next various automatic actions take place. The walls of the pharynx contract and propel the bolus of food toward the esophagus. The soft palate rises so that the bolus does not enter the nasal passages (2). The epiglottis, a cartilage that acts as a valve, covers the larynx so that the bolus does not enter the air passages (3). When the bolus reaches the esophagus, a series of sequential muscular contractions of the walls of this organ convey the bolus downward (4 and 5) until it is finally propelled into the stomach (6).

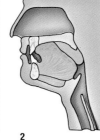
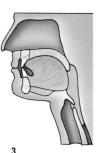

1   2   3

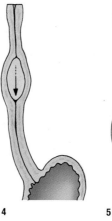
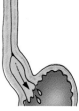

4   5   6

Even when we do not eat anything, the act of swallowing is repeated incessantly. On average, we swallow saliva around 70 times an hour while we are awake and around 10 times an hour while we are asleep.

PROJECTION OF THE STOMACH ONTO THE SURFACE OF THE BODY

The stomach is a hollow organ with muscular walls. It has two orifices. The upper one, named the **cardia**, prevents the contents of the stomach from flowing back into the esophagus. The lower one, the **pylorus**, acts as a valve, remaining shut until the food is ready to continue its journey and then opening to allow the food to pass into the duodenum.

SECTION OF THE STOMACH
ANTERIOR VIEW

esophagus

cardia

lesser curvature

duodenum

pylorus

fundus of stomach

body of stomach

greater curvature

gastric folds

pyloric antrum

mucous membrane

submucosa

serosa

muscle layer

The stomach receives the bolus of food, and thanks to the powerful, peristaltic movement of the stomach walls, mixes it with **gastric juice**. This submits the food to the chemical effects of gastric juice's two main components: very strong **hydrochloric acid** and the enzyme **pepsin**.

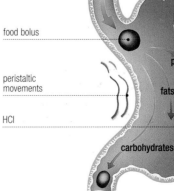

food bolus

peristaltic movements

HCl

pepsin

proteins

fats

carbohydrates

Introduction

The cell

The human body

The locomotive system

The digestive system

The respiratory system

The circulatory system

Blood

Lymph

The nervous system

The senses

The urinary system

The reproductive system

Human reproduction

The endocrine system

The immune system

Alphabetical index

SMALL INTESTINE

The small intestine is where the main effects in the digestive process take place. Here food is submitted to the action of enzymes that come from the liver, the pancreas, and the mucous membrane of the small intestine itself. These enzymes break down the food into more basic elements. It is a tube about 23 to 26 feet (7 to 8 m) long that, although continuous, consists of three sections:

• the **duodenum** is the section located at the exit to the stomach, is about 10 to 12 inches (25 to 30 cm) long, and contains secretions from the pancreas and bile salts from the liver;

• the **jejunum** is located in the upper part of the abdominal cavity and is about 10 feet (3 m) long;

• the **ileum** is located in the lower part of the abdominal cavity, is 10 to 13 feet (3 to 4 m) long, and ends in the large intestine.

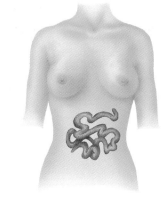

DUODENUM
ANTERIOR VIEW

duodenal chamber

upper horizontal section

minor duodenal papilla

pylorus

stomach

common bile duct

secondary pancreatic duct

main pancreatic duct

descending section

jejunum

ascending section

major duodenal papilla or Vater's duct

lower horizontal section

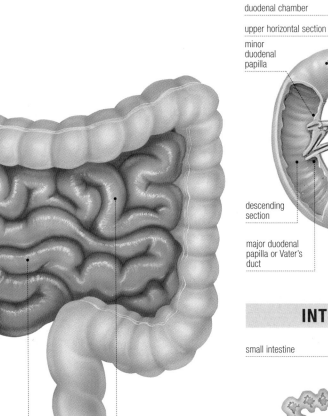

íleum

jejunum

ANTERIOR VIEW OF THE SMALL INTESTINE, SURROUNDED BY THE LARGE INTESTINE

INTESTINAL MOVEMENT

small intestine

1

2

large intestine

3

ileocecal valve

The walls of the small intestine contract automatically for different purposes. Rhythmic movements of different sections compact and crush the food (*1*). Contractions of each of pair of rings in opposite directions act to mix it well (*2*). Sequential peristaltic movements propel the food in the direction of the large intestine (*3*).

The pancreas is a **gland** attached to the digestive tract. It produces a secretion that is rich in enzymes, whose task is to break down food. The pancreas also forms part of the endocrine system because it produces important hormones such as insulin.

It is a long, cone-shaped organ that is located transversely in the upper part of the abdomen. The largest part, the head, is attached to the duodenum, into which it empties digestive secretions.

Introduction

The cell

The human body

The locomotive system

The digestive system

The respiratory system

The circulatory system

Blood

Lymph

The nervous system

The senses

The urinary system

The reproductive system

Human reproduction

The endocrine system

The immune system

Alphabetical index

PARTIAL SECTION OF THE PANCREAS AND DUODENUM

ANTERIOR VIEW

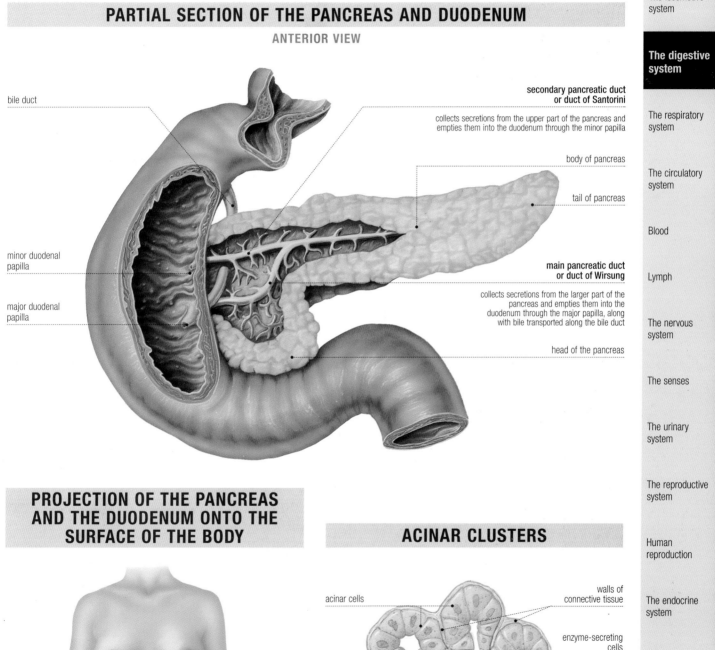

bile duct

secondary pancreatic duct or duct of Santorini

collects secretions from the upper part of the pancreas and empties them into the duodenum through the minor papilla

body of pancreas

tail of pancreas

minor duodenal papilla

major duodenal papilla

main pancreatic duct or duct of Wirsung

collects secretions from the larger part of the pancreas and empties them into the duodenum through the major papilla, along with bile transported along the bile duct

head of the pancreas

PROJECTION OF THE PANCREAS AND THE DUODENUM ONTO THE SURFACE OF THE BODY

ACINAR CLUSTERS

acinar cells

walls of connective tissue

enzyme-secreting cells

canaliculi

LIVER

The liver is a gland linked to the digestive system. In addition to carrying out various functions that are essential for metabolism, it produces bile, a vital secretion required for digesting fats. Bile is stored in the gallbladder until it is sufficiently concentrated and is then emptied after each meal into the small intestine by means of bile ducts.

PROJECTION OF THE LIVER ONTO THE SURFACE OF THE BODY

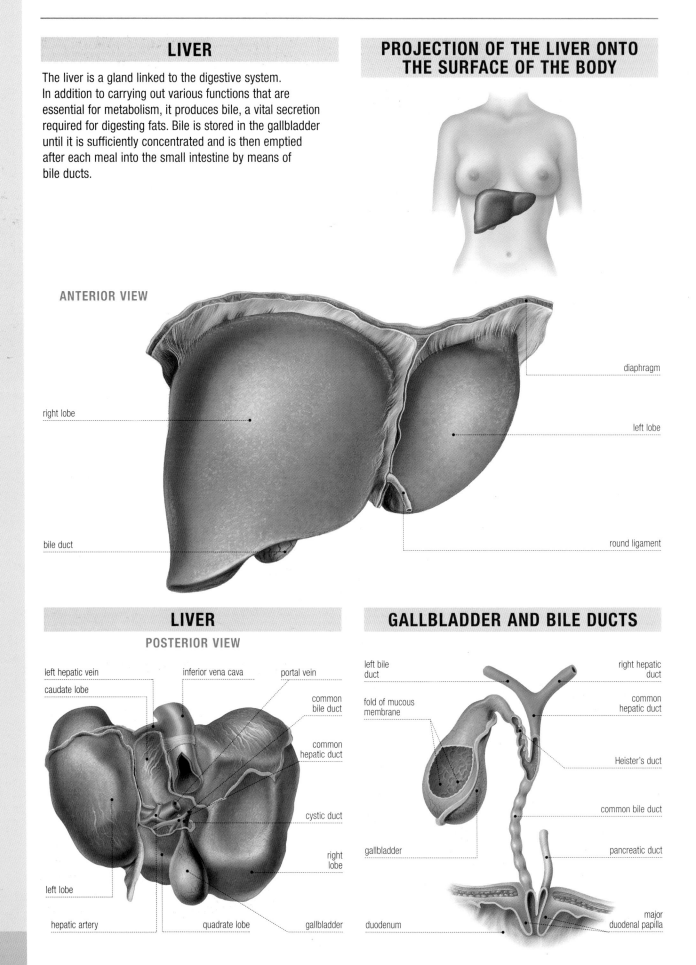

ANTERIOR VIEW

right lobe

bile duct

diaphragm

left lobe

round ligament

LIVER

POSTERIOR VIEW

left hepatic vein

caudate lobe

inferior vena cava

portal vein

common bile duct

common hepatic duct

cystic duct

right lobe

left lobe

hepatic artery

quadrate lobe

gallbladder

GALLBLADDER AND BILE DUCTS

left bile duct

fold of mucous membrane

gallbladder

duodenum

right hepatic duct

common hepatic duct

Heister's duct

common bile duct

pancreatic duct

major duodenal papilla

LARGE INTESTINE

The large intestine is the **final section** of the digestive tract. The residue from the digestive process is temporarily stored there while it is converted into **feces** that are then expelled from the body. It is a tube about 5 to 6 feet (1.5 to 1.8 m) long, consisting of three sections:

• the **cecum** is located in the lower, right-hand part of the abdomen, which opens into the small intestine;

• the **colon**, the longest section, is arranged around the interior of the abdominal cavity and divided into four sections: ascending, transverse, descending, and sigmoid;

• the **rectum** ends at the anus.

PROJECTION OF THE LARGE INTESTINE ONTO THE SURFACE OF THE BODY

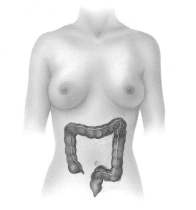

ANTERIOR VIEW OF THE LARGE INTESTINE

haustra

transverse colon

descending colon

ascending colon

ileum

sigmoid colon

cecum

rectum

vermiform appendix

FRONTAL SECTION OF CECUM

teniae coli

haustra

ascending colon

iliocecal sphincter

ileum

teniae coli

cecum

vermiform appendix

opening of vermiform appendix

FRONTAL SECTION OF RECTUM

sigmoid colon

rectal chamber

rectal valves

internal anal sphincter

external anal sphincter

anal duct

Introduction

The cell

The human body

The locomotive system

The digestive system

The respiratory system

The circulatory system

Blood

Lymph

The nervous system

The senses

The urinary system

The reproductive system

Human reproduction

The endocrine system

The immune system

Alphabetical index

THE RESPIRATORY SYSTEM

The respiratory apparatus is responsible for maintaining a constant **exchange of gases** between the body and the air around it. This is an essential function that allows us to **take in oxygen**, used by all the tissues as fuel for the production of energy, and to **eliminate carbon dioxide**, which is generated as a residue of this process and which can be toxic if it builds up in the body.

ORGANS OF THE RESPIRATORY APPARATUS

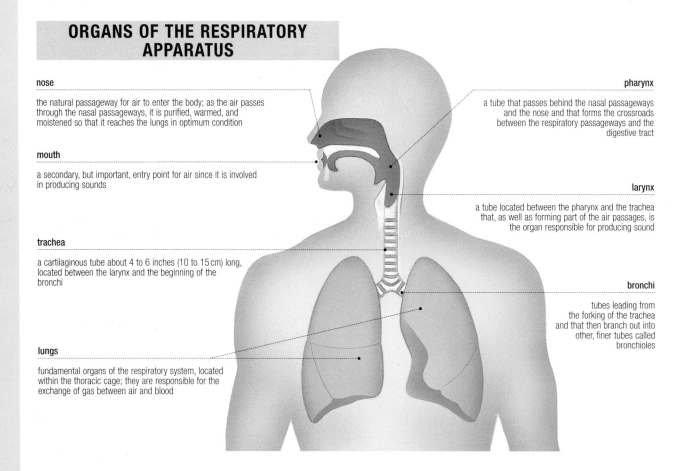

nose

the natural passageway for air to enter the body; as the air passes through the nasal passageways, it is purified, warmed, and moistened so that it reaches the lungs in optimum condition

mouth

a secondary, but important, entry point for air since it is involved in producing sounds

trachea

a cartilaginous tube about 4 to 6 inches (10 to 15 cm) long, located between the larynx and the beginning of the bronchi

lungs

fundamental organs of the respiratory system, located within the thoracic cage; they are responsible for the exchange of gas between air and blood

pharynx

a tube that passes behind the nasal passageways and the nose and that forms the crossroads between the respiratory passageways and the digestive tract

larynx

a tube located between the pharynx and the trachea that, as well as forming part of the air passages, is the organ responsible for producing sound

bronchi

tubes leading from the forking of the trachea and that then branch out into other, finer tubes called bronchioles

MECHANISM OF BREATHING

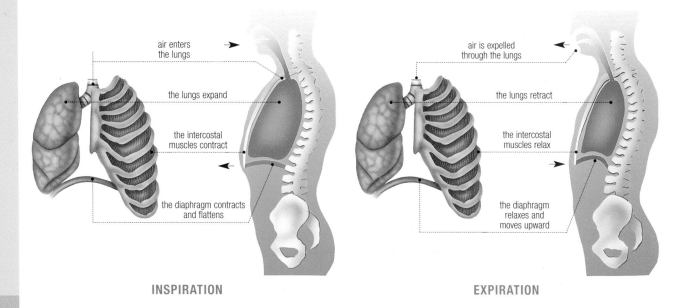

air enters the lungs

the lungs expand

the intercostal muscles contract

the diaphragm contracts and flattens

air is expelled through the lungs

the lungs retract

the intercostal muscles relax

the diaphragm relaxes and moves upward

INSPIRATION

EXPIRATION

BONES AND CARTILAGE OF THE NASAL PYRAMID

frontal bone

maxilla bone

nasal bone

upper nasal cartilage

septal cartilage

lateral crus of greater alar cartilage

medial crus of greater alar cartilage

nasal wall cartilage

PARANASAL SINUSES

frontal sinus

ethmoidal sinus

sphenoidal sinus

maxillary sinus

The paranasal sinuses are **cavities** present in some bones of the skull. They are full of air and covered by a mucous membrane similar to the nose. They are directly connected to the nasal passageways and act as a **sounding box** when speaking.

PHARYNX

POSTERIOR VIEW

The pharynx is the **tube** that starts at the base of the nasal passageways and goes down behind the mouth to the larynx and esophagus. It represents a **common opening for air and food**. Therefore, it forms part of both the respiratory and digestive systems. At the point where the pharynx and larynx meet is cartilage in the form of a flap, called the **epiglottis**. During swallowing, this bends backward and covers the entrance to the air passages.

LATERAL SECTION OF THE NASAL CAVITY

frontal bone

frontal sinus

nasal bone

middle nasal concha

inferior nasal concha

nasal vestibule nasal

olfactory membrane

superior nasal concha

sphenoidal bone

superior maxillary bone

pharynx

The nose is the **natural entranceway for air** into the lungs. It is located in the center of the face and consists of a pyramid-shaped protuberance, with the **nasal orifices** at its base. Inside it, there are two large areas, the **nasal cavities**, separated by a dividing wall.

LATERAL SECTION OF PHARYNX

choanae

nasal septum

sphenoidal sinus

pharyngeal tonsil

opening of auditory (eustachian) tube

superior maxillary bone

soft palate

nasopharynx

tongue

palatine tonsil

lingual tonsil

epiglottis

oropharynx

laryngopharynx

vocal chords

larynx

trachea

esophagus

nasal pharynx

oral pharynx

epiglottis

laryngeal pharynx

larynx

Introduction

The cell

The human body

The locomotive system

The digestive system

The respiratory system

The circulatory system

Blood

Lymph

The nervous system

The senses

The urinary system

The reproductive system

Human reproduction

The endocrine system

The immune system

Alphabetical index

FRONTAL SECTION OF THE LARYNX

The larynx is a **tube** made up of various pieces of **cartilage linked** to one another, joining the pharynx with the trachea. It is therefore part of the pathway that air passes through on its way to the lungs during inspiration and in the opposite direction during expiration. It is also the **organ of phonation** since it is in the larynx that the **vocal cords** are found.

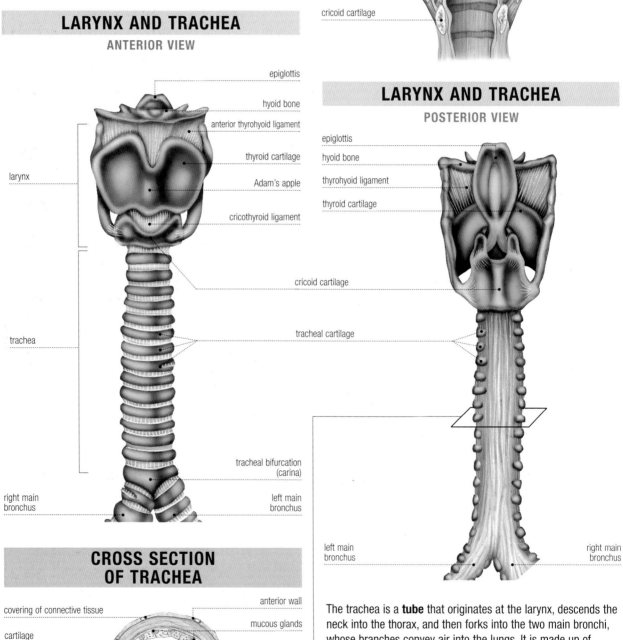

epiglottis
false vocal cords
hyoid bones
Morgani's ventricles
thyroid cartilage
vocal cords
cricoid cartilage

LARYNX AND TRACHEA
ANTERIOR VIEW

epiglottis
hyoid bone
anterior thyrohyoid ligament
thyroid cartilage
Adam's apple
cricothyroid ligament
larynx
trachea
cricoid cartilage
tracheal cartilage
tracheal bifurcation (carina)
right main bronchus
left main bronchus

LARYNX AND TRACHEA
POSTERIOR VIEW

epiglottis
hyoid bone
thyrohyoid ligament
thyroid cartilage
cricoid cartilage
tracheal cartilage
left main bronchus
right main bronchus

CROSS SECTION OF TRACHEA

covering of connective tissue
cartilage
epithelium
tracheal muscle
anterior wall
mucous glands
tracheal passageway
posterior wall
esophageal muscle

The trachea is a **tube** that originates at the larynx, descends the neck into the thorax, and then forks into the two main bronchi, whose branches convey air into the lungs. It is made up of around 15 or 20 horseshoe-shaped pieces of **cartilage**. These do not meet at the back but almost fully enclose the circumference of the tube. The posterior part that is not covered by cartilage is covered by membrane and consists of connective tissue and muscle.

BRONCHIAL TREE
ANTERIOR VIEW IDENTIFIED BY COLOR

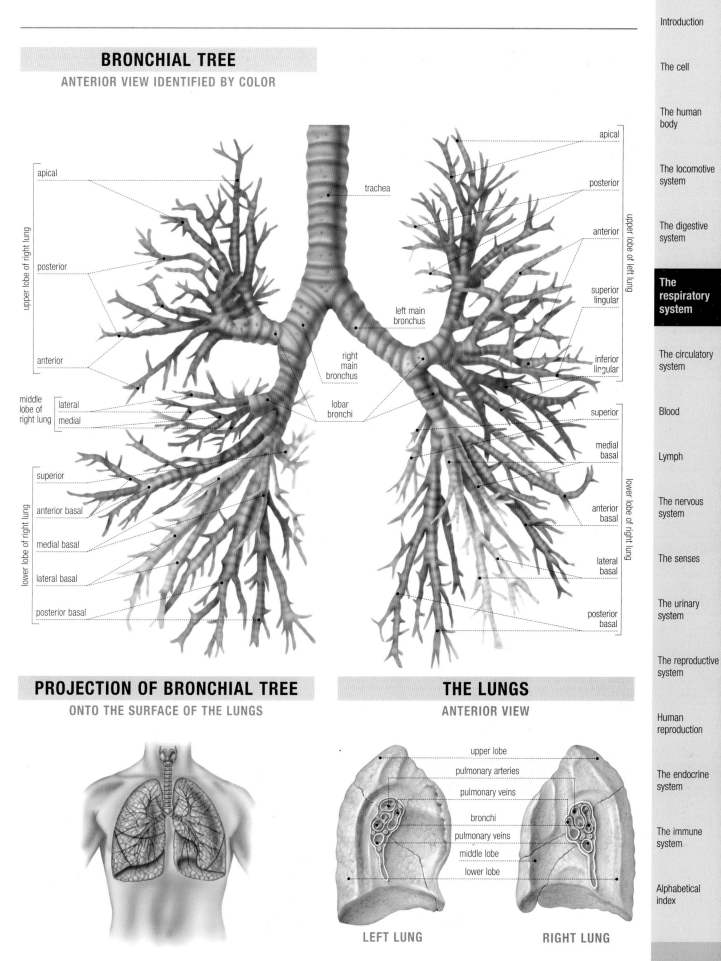

apical

trachea

posterior

anterior

upper lobe of right lung

apical

posterior

anterior

upper lobe of left lung

superior lingular

inferior lingular

left main bronchus

right main bronchus

lobar bronchi

middle lobe of right lung
- lateral
- medial

superior

medial basal

anterior basal

lateral basal

posterior basal

lower lobe of right lung

superior

anterior basal

medial basal

lateral basal

posterior basal

lower lobe of right lung

PROJECTION OF BRONCHIAL TREE
ONTO THE SURFACE OF THE LUNGS

THE LUNGS
ANTERIOR VIEW

upper lobe

pulmonary arteries

pulmonary veins

bronchi

pulmonary veins

middle lobe

lower lobe

LEFT LUNG

RIGHT LUNG

Introduction

The cell

The human body

The locomotive system

The digestive system

The respiratory system

The circulatory system

Blood

Lymph

The nervous system

The senses

The urinary system

The reproductive system

Human reproduction

The endocrine system

The immune system

Alphabetical index

THE CIRCULATORY SYSTEM

The circulatory system consists of a **network of blood vessels** that, with the rhythmic pumping of the heart, conveys **blood** around the body incessantly. The blood transports elements that all the tissues of the body need to maintain their vital activity. It also collects the residues of cellular metabolism, taking them to the organs responsible for getting rid of them.

DIAGRAM OF THE CIRCULATORY SYSTEM

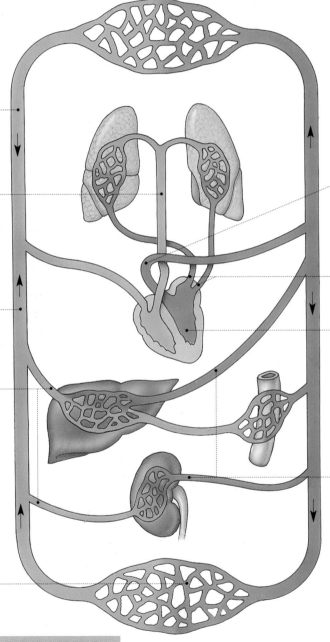

superior vena cava

conveys blood that is low in oxygen from the veins in the upper part of the body to the heart

pulmonary artery

rreceives blood low in oxygen, pumped by the heart, and conveys it to the lungs

inferior vena cava

conveys blood that is low in oxygen from the veins in the lower part of the body to the heart

veins

convey blood that is low in oxygen toward the heart, via the vena cava

capillaries

these are the narrowest blood vessels; through their fine walls the interchange between blood and tissue takes place

aorta

this is the main artery in the body; it receives blood that is high in oxygen, pumped by the heart, and distributes it along its branches to reach every part of the body

pulmonary veins

carry blood that has been oxygenated in the lungs to the heart

heart

this is the central motor of the circulatory system; with each heartbeat, it pumps blood through the arteries; the blood goes around the body and then returns to the heart via the veins

arteries

carry oxygenated blood from the heart to the different tissues of the body

BLOOD VESSELS

The heart pumps blood that is high in oxygen to the aorta. This is a large **artery** with numerous branches that, like the branches of a tree, divide repeatedly. These turn into smaller blood vessels called **arterioles**, which join up with very fine vessels called **capillaries**. Capillary walls consist of a single layer of cells. These walls are so thin that they allow exchanges of chemicals between the blood and the body tissues. Next, the **capillaries** join with **venules**, which join together to become bigger and bigger **veins**. These carry blood that is low in oxygen and loaded with waste residues in the direction of the heart.

HEART
ANTERIOR VIEW

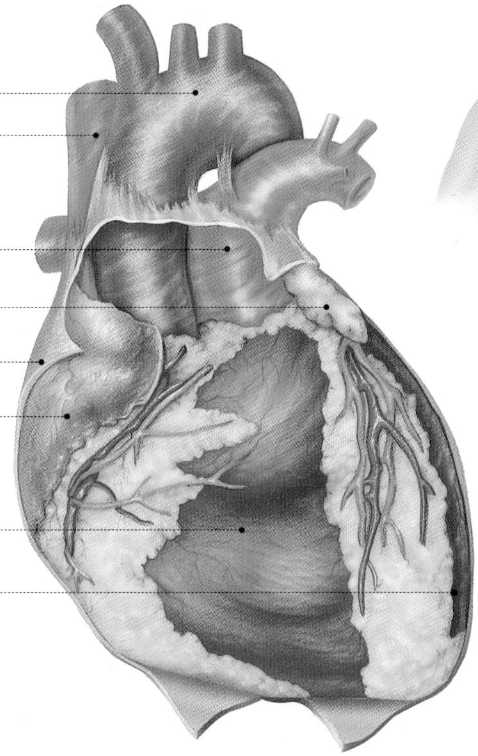

aorta

superior vena cava

pulmonary artery

left atrium

pericardium (section)

right atrium

right ventricle

left ventricle

PROJECTION OF THE HEART ONTO THE SURFACE OF THE BODY

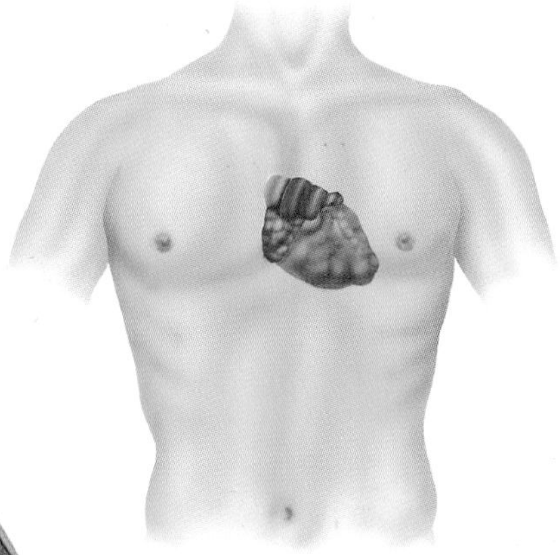

The heart, the motor of the circulatory system, is an organ the size of a closed fist. It is located in the center of the thorax between the two lungs in a space called the **mediastinum**. It is covered in a double serous membrane, called the **pericardium**. The heart is shaped like an irregular cone positioned obliquely, with the base pointing up and to the right and the tip pointing down and to the left.

HEART
POSTERIOR VIEW

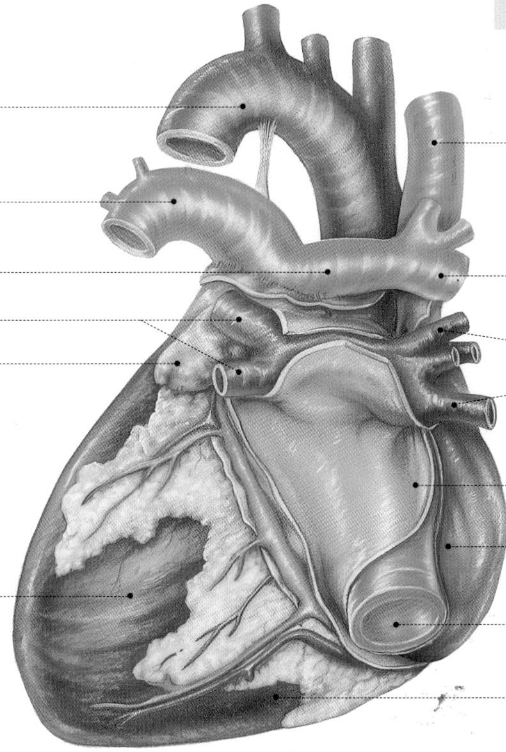

aorta

left pulmonary artery

trunk of pulmonary artery

left pulmonary veins

left atrium

left ventricle

superior vena cava

right pulmonary artery

right pulmonary veins

pericardium (section)

right atrium

inferior vena cava

right ventricle

Introduction

The cell

The human body

The locomotive system

The digestive system

The respiratory systxem

The circulatory system

Blood

Lymph

The nervous system

The senses

The urinary system

The reproductive system

Human reproduction

The endocrine system

The immune system.

Alphabetical index

LONGITUDINAL SECTION OF THE HEART

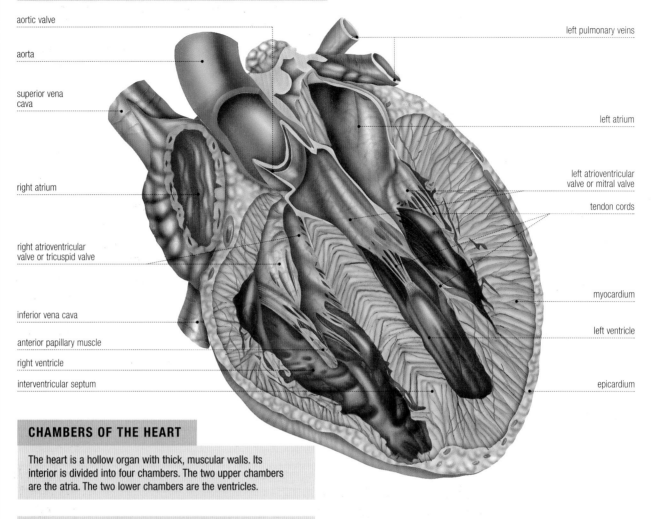

aortic valve

aorta

superior vena cava

right atrium

right atrioventricular valve or tricuspid valve

inferior vena cava

anterior papillary muscle

right ventricle

interventricular septum

left pulmonary veins

left atrium

left atrioventricular valve or mitral valve

tendon cords

myocardium

left ventricle

epicardium

CHAMBERS OF THE HEART

The heart is a hollow organ with thick, muscular walls. Its interior is divided into four chambers. The two upper chambers are the atria. The two lower chambers are the ventricles.

CARDIAC VALVES

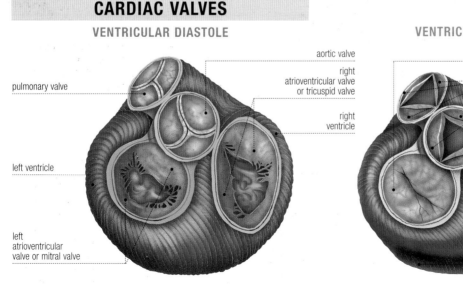

VENTRICULAR DIASTOLE

pulmonary valve

left ventricle

left atrioventricular valve or mitral valve

aortic valve

right atrioventricular valve or tricuspid valve

right ventricle

VENTRICULAR SYSTOLE

left atrioventricular valve or mitral valve

pulmonary valve

aortic valve

right atrioventricular valve or tricuspid valve

right ventricle

left ventricle

Blood circulates through the heart **in a single direction**. It travels from each atrium to the ventricle on the same side and from each ventricle into the artery that connects to it. The pulmonary artery connects on the right side, the aorta on the left side. This **unidirectional** flow is possible thanks to four **valves** that are synchronized with different phases of the heartbeat. These valves allow the blood to pass from one section to the next while preventing it from flowing back again. There are two **atrioventricular valves**, one on the right (the tricuspid valve) and one on the left (the mitral valve). There are two **semilunar valves**, one located between the right ventricle and the pulmonary artery (the pulmonary valve) and the other located between the left ventricle and the aorta (the aortic valve).

CORONARY BLOOD VESSELS
ANTERIOR AND POSTERIOR VIEWS

The heart has its own **blood supply** that is provided by a network of blood vessels that surround it like a crown. The two main coronary arteries, the left and the right, spread out from the aorta. Their many branches take oxygenated blood to all areas of the heart. After irrigating the cardiac tissue, the blood, now low in oxygen, is conveyed along a network of small veins that join together to make increasingly large veins that reach the coronary sinus, a tube that connects with the right atrium.

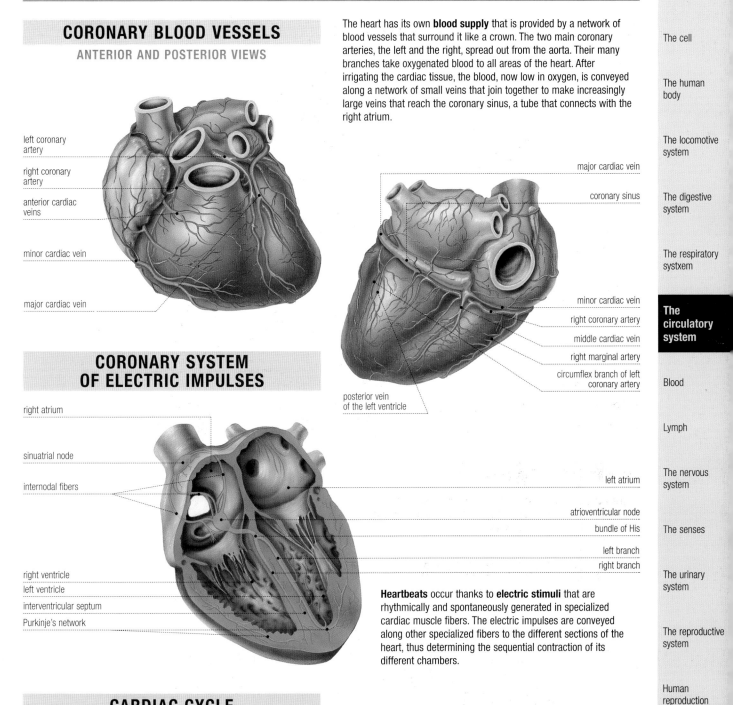

left coronary artery

right coronary artery

anterior cardiac veins

minor cardiac vein

major cardiac vein

major cardiac vein

coronary sinus

minor cardiac vein

right coronary artery

middle cardiac vein

right marginal artery

circumflex branch of left coronary artery

posterior vein of the left ventricle

CORONARY SYSTEM OF ELECTRIC IMPULSES

right atrium

sinuatrial node

internodal fibers

left atrium

atrioventricular node

bundle of His

left branch

right branch

right ventricle

left ventricle

interventricular septum

Purkinje's network

Heartbeats occur thanks to **electric stimuli** that are rhythmically and spontaneously generated in specialized cardiac muscle fibers. The electric impulses are conveyed along other specialized fibers to the different sections of the heart, thus determining the sequential contraction of its different chambers.

CARDIAC CYCLE

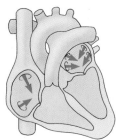

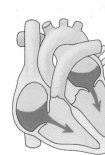

DIASTOLE

ATRIAL SYSTOLE

VENTRICULAR SYSTOLE

In each heartbeat, the **dilation** (diastole) and the **contraction** (systole) of each cardiac chamber take place synchronously. The blood passes from each atrium to the ventricle on the respective side and from there to the respective artery in a cycle that repeats itself endlessly.

Introduction

The cell

The human body

The locomotive system

The digestive system

The respiratory systxem

The circulatory system

Blood

Lymph

The nervous system

The senses

The urinary system

The reproductive system

Human reproduction

The endocrine system

The immune system

Alphabetical index

MAIN ARTERIES OF THE ORGANISM

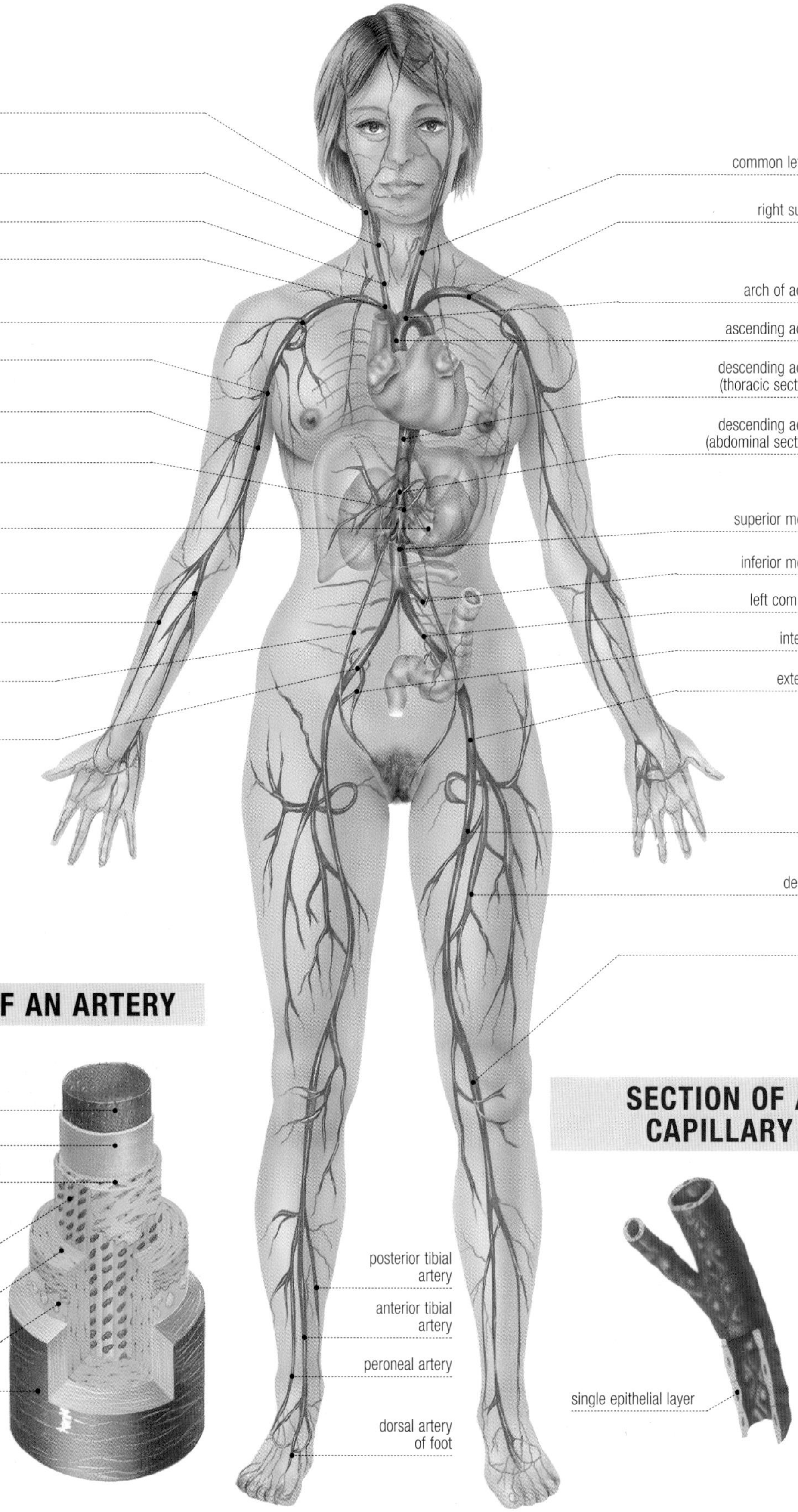

external carotid artery

internal carotid artery

common right carotid artery

brachiocephalic trunk

right subclavian artery

axillary artery

femoral artery

celiac trunk

renal artery

ulnar artery

radial artery

ovarian/testicular artery

right common iliac artery

common left carotid artery

right subclavian artery

arch of aorta

ascending aorta

descending aorta
(thoracic section)

descending aorta
(abdominal section)

aorta

superior mesenteric artery

inferior mesenteric artery

left common iliac artery

internal iliac artery

external iliac artery

femoral artery

deep femoral vein

popliteal artery

posterior tibial
artery

anterior tibial
artery

peroneal artery

dorsal artery
of foot

SECTION OF AN ARTERY

tunica intima
- endothelium
- basal membrane
- lamina of elastic fibers

tunica media
- internal elastic membrane
- transverse elastic muscle fibers
- external elastic membrane

tunica adventitia

SECTION OF A CAPILLARY

single epithelial layer

MAIN VEINS OF THE ORGANISM

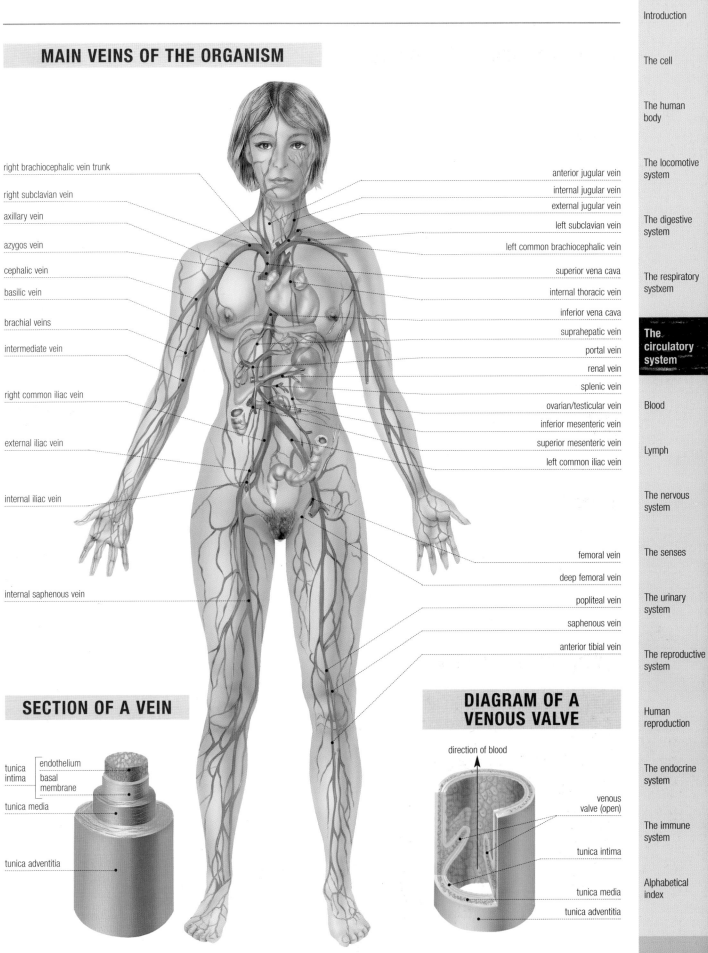

right brachiocephalic vein trunk

right subclavian vein

axillary vein

azygos vein

cephalic vein

basilic vein

brachial veins

intermediate vein

right common iliac vein

external iliac vein

internal iliac vein

internal saphenous vein

anterior jugular vein

internal jugular vein

external jugular vein

left subclavian vein

left common brachiocephalic vein

superior vena cava

internal thoracic vein

inferior vena cava

suprahepatic vein

portal vein

renal vein

splenic vein

ovarian/testicular vein

inferior mesenteric vein

superior mesenteric vein

left common iliac vein

femoral vein

deep femoral vein

popliteal vein

saphenous vein

anterior tibial vein

SECTION OF A VEIN

tunica intima — endothelium / basal membrane

tunica media

tunica adventitia

DIAGRAM OF A VENOUS VALVE

direction of blood

venous valve (open)

tunica intima

tunica media

tunica adventitia

Introduction

The cell

The human body

The locomotive system

The digestive system

The respiratory systxem

The circulatory system

Blood

Lymph

The nervous system

The senses

The urinary system

The reproductive system

Human reproduction

The endocrine system

The immune system

Alphabetical index

BLOOD

Blood is a red-colored, viscous **liquid** that travels around the circulatory system continuously. It transports **oxygen**, **nutrients**, and other elements that cells require for metabolism to all the tissues of the body. It takes **toxic waste products** to the organs responsible for getting rid of them.

THE COMPOSITION OF BLOOD

Blood consists of **plasma**, a yellowish liquid made up mostly of water that contains many substances dissolved in it and that carries numerous **blood cells** in suspension.

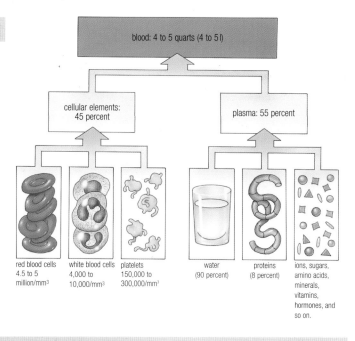

blood: 4 to 5 quarts (4 to 5 l)

cellular elements: 45 percent

plasma: 55 percent

red blood cells
4.5 to 5 million/mm³

white blood cells
4,000 to 10,000/mm³

platelets
150,000 to 300,000/mm³

water
(90 percent)

proteins
(8 percent)

ions, sugars, amino acids, minerals, vitamins, hormones, and so on.

BLOOD CELL FORMATION

New blood corpuscles are cells produced to replace cells that grow old and are therefore destroyed. Each day, thousands and thousands of millions of red and white blood cells and platelets are produced. This process is called hematopoiesis. It takes place basically in the bone marrow, starting with communal primary cells called pluripotent stem cells. These are capable of reproducing themselves and of creating different monopotent mother cells, which mature to create the different blood cells.

BLOOD CELLS

Different types of cells float in blood plasma, each one of which has a specific function:

• **red blood cells**, also called erythrocytes, are responsible for transporting oxygen from the lungs to the body tissue and transporting in the opposite direction the carbon dioxide that results from metabolism;

• **white blood cells**, also called leukocytes, take different forms and are part of the immune system, protecting the organism from infections;

• **platelets**, also called thrombocytes, are involved in coagulation to stop bleeding.

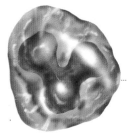

red blood cells

(anterior view)

(lateral view)

platelet

white blood cell

BONE MARROW

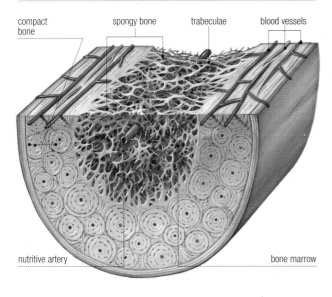

compact bone · spongy bone · trabeculae · blood vessels

nutritive artery · bone marrow

LOCATION OF ACTIVE BONE MARROW IN ADULTS

Blood cells are mostly produced in the bone marrow, a special tissue found **inside bones**. In newborn babies, all the bones of the skeleton contain active bone marrow. As the child grows, and particularly from adolescence onward, much of the bone marrow is replaced by fatty tissue.

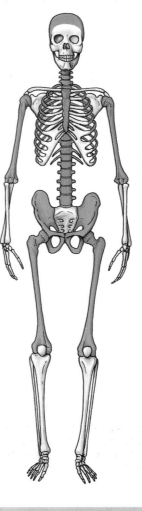

SPLEEN

ANTEROMEDIAL VIEW

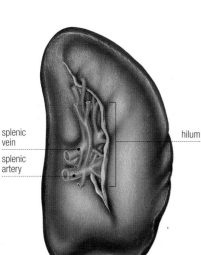

splenic vein

splenic artery

hilum

SECTION

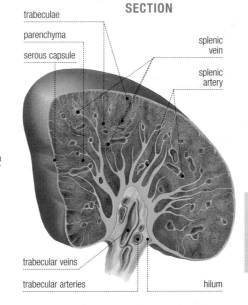

trabeculae
parenchyma
serous capsule

splenic vein
splenic artery

trabecular veins
trabecular arteries
hilum

The spleen is an organ located in the upper left part of the abdomen. During life in the womb, it produces all types of blood cells. After birth, it produces only some white blood cells. Its main function is to destroy **old blood cells**. However, it is also part of the immune system since it acts as a **filter** for germs and impurities in the blood that circulates through it.

MICROSCOPIC STRUCTURE OF THE SPLEEN

malpighian cell
white pulp
red pulp
vein sinuses
Billroth's cords
trabeculum
trabecular artery
trabecular vein

PROJECTION OF THE SPLEEN ONTO THE SURFACE OF THE BODY

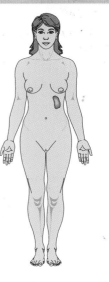

Introduction

The cell

The human body

The locomotive system

The digestive system

The respiratory system

The circulatory system

Blood

Lymph

The nervous system

The senses

The urinary system

The reproductive system

Human reproduction

The endocrine system

The immune system

Alphabetical index

LYMPH

The lymphatic system consists of an intricate network of vessels, called **lymphatic ducts**. These drain away the liquid that fills the spaces between **cells** and the miniscule particles present in this liquid. It conveys this liquid, called **lymph**, toward the circulatory system to incorporate it into the bloodstream. On its way, it passes through nodular formations, the **lymph nodes**, that store large quantities of white blood cells and **act as a filter** for germs and impurities.

RELATIONSHIP BETWEEN THE LYMPHATIC SYSTEM AND THE CIRCULATORY SYSTEM

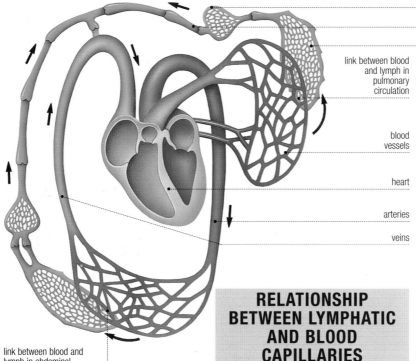

lymph node
lymphatic vessels
lymphatic capillaries
link between blood and lymph in pulmonary circulation
blood vessels
heart
arteries
veins
link between blood and lymph in abdominal circulation

The main task of the lymphatic system is to **collect** liquid plasma that passes from blood capillaries to intercellular spaces in tissue, draining these small spaces between cells so that they do not flood. It does so through a complex network of ducts and channels that finally empty out into the venous system, where the excess liquid is reincorporated into the circulatory system.

SCHEMATIC CROSS SECTION OF A LYMPHATIC CAPILLARY

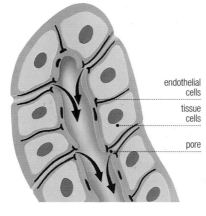

endothelial cells
tissue cells
pore

RELATIONSHIP BETWEEN LYMPHATIC AND BLOOD CAPILLARIES

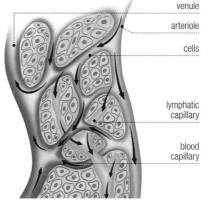

venule
arteriole
cells
lymphatic capillary
blood capillary

Lymphatic capillaries are present in all body tissues. They are **very fine vessels**, with one end closed off, and whose walls are formed of a single layer of endothelial cells. They absorb excess liquid, proteins, germs, and all types of foreign particles present around them through **pores** between the cells.

THE PASSAGE OF WHITE BLOOD CELLS INTO THE LYMPHATIC SYSTEM

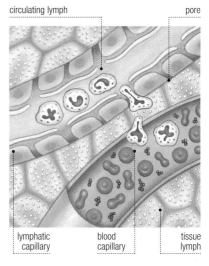

circulating lymph
pore
lymphatic capillary
blood capillary
tissue lymph

Numerous white blood cells are present in lymph. Their purpose is to **defend the organism**. Many of the white blood cells pass from blood vessels into lymphatic vessels.

LYMPHATIC VESSELS

EXTERIOR VIEW

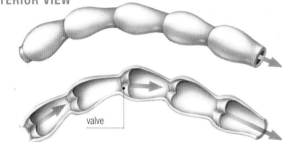

valve

LONGITUDINAL SECTION

Lymphatic vessels represent a **continuation of capillaries**, which increase progressively in diameter, joining together to form other, thicker vessels. Inside they have **valves** that permit the lymph to travel in just one direction, preventing it from flowing backward and ensuring that it circulates in the correct direction.

LYMPH NODE

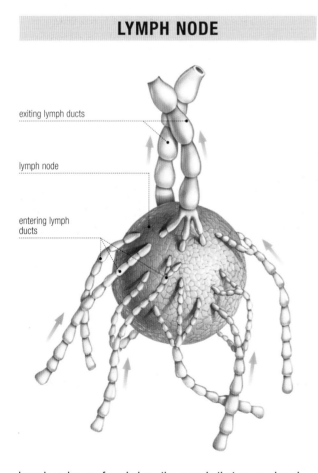

exiting lymph ducts

lymph node

entering lymph ducts

Lymph nodes are found along the vessels that convey lymph. They are **globular formations** that, under normal conditions, are never more than 0.8 inch (2 cm) in diameter. They consist of a fibrous external capsule from which spread various dividing walls. These separate different sections inside the side, where there is an accumulation of **lymphoid tissue** that stores lots of white blood cells for **defending** the organism against foreign bodies.

SCHEMATIC REPRESENTATION OF THE LYMPHATIC SYSTEM

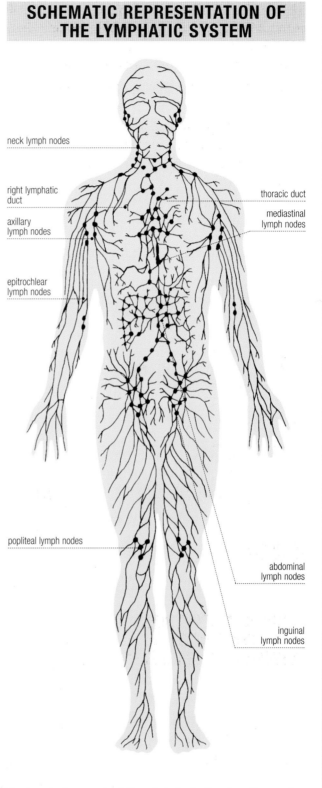

neck lymph nodes

right lymphatic duct

thoracic duct

axillary lymph nodes

mediastinal lymph nodes

epitrochlear lymph nodes

popliteal lymph nodes

abdominal lymph nodes

inguinal lymph nodes

The lymphatic vessels of the whole body **flow together** and finally empty their contents into two large channels, the **thoracic duct** and the **right lymphatic duct**. These channels empty respectively into the left and right subclavian veins, which in turn empty into the superior vena cava. In this way, lymph reaches the blood circulatory system.

Introduction

The cell

The human body

The locomotive system

The digestive system

The respiratory system

The circulatory system

Blood

Lymph

The nervous system

The senses

The urinary system

The reproductive system

Human reproduction

The endocrine system

The immune system

Alphabetical index

THE NERVOUS SYSTEM

The nervous system consists of the organs that make up the brain, the spinal cord, and a network of nerves that spread to every part of the body. It directs our **voluntary actions**, regulating the **automatic functions** of the organism, is responsible for the **relationship** we have with our external environment, and is the seat of **intellectual activity**.

COMPONENTS OF THE NERVOUS SYSTEM

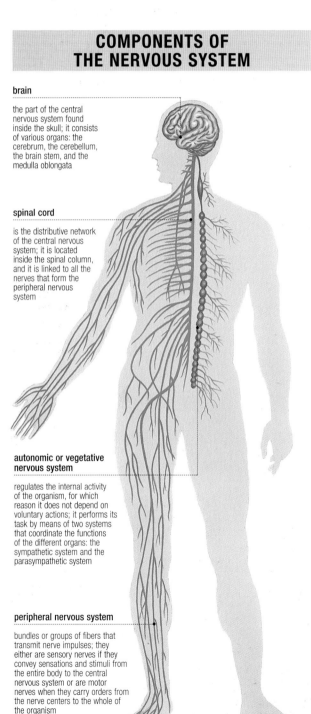

brain

the part of the central nervous system found inside the skull; it consists of various organs: the cerebrum, the cerebellum, the brain stem, and the medulla oblongata

spinal cord

is the distributive network of the central nervous system; it is located inside the spinal column, and it is linked to all the nerves that form the peripheral nervous system

autonomic or vegetative nervous system

regulates the internal activity of the organism, for which reason it does not depend on voluntary actions; it performs its task by means of two systems that coordinate the functions of the different organs: the sympathetic system and the parasympathetic system

peripheral nervous system

bundles or groups of fibers that transmit nerve impulses; they either are sensory nerves if they convey sensations and stimuli from the entire body to the central nervous system or are motor nerves when they carry orders from the nerve centers to the whole of the organism

STRUCTURE OF A NEURON

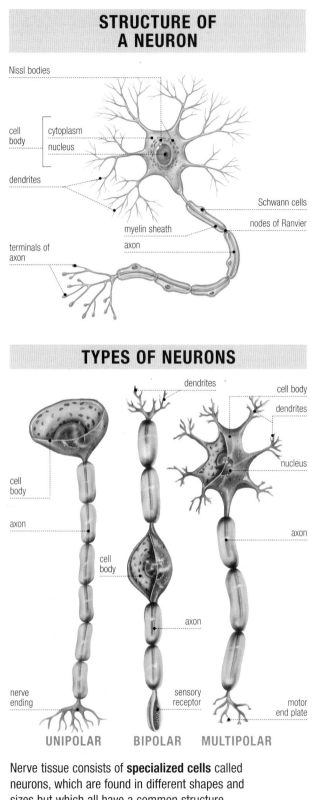

Nissl bodies
cell body
cytoplasm
nucleus
dendrites
Schwann cells
nodes of Ranvier
myelin sheath
axon
terminals of axon

TYPES OF NEURONS

dendrites
cell body
dendrites
nucleus
cell body
axon
cell body
axon
nerve ending
sensory receptor
motor end plate
axon

UNIPOLAR   **BIPOLAR**   **MULTIPOLAR**

Nerve tissue consists of **specialized cells** called neurons, which are found in different shapes and sizes but which all have a common structure. Each neuron has a cell body from which emerge two extensions. The **dendrites** are short, treelike branches along which impulses travel from other nerve cells. The **axon** is a single, long extension along which impulses are transmitted to other cells or to the body tissue.

LONGITUDINAL SECTION OF BRAIN

cerebrum

pons

medulla oblongata

spinal cord

cerebellum

The brain is the part of the nervous system that consists of the structures contained in the skull:

• the **cerebrum** is the largest and most important organ since it controls all voluntary activity and a great deal of involuntary activity of the body as well as being the seat of mental processes;

• the **brain stem** consists of the pons and the medulla oblongata where vital functions are regulated and where most of the nuclei of most of the cranial nerves originate;

• the **cerebellum** is involved in controlling balance and modifying body movements.

INTERIOR VIEW OF THE BRAIN

FRONTAL POLE

cerebrum

olfactory nerve (I cranial pair)

olfactory bulb

olfactory tract

ophthalmic groove

hypophysis

mamillary body

pons

medulla oblongata

cerebellum

occipital lobe

optic nerve
(II cranial pair)

common ocular motor nerve
(III cranial pair)

ophthalmic nerve

superior maxillary nerve

trochlear nerve
(IV cranial pair)

inferior maxillary nerve

trigeminal nerve
(V cranial pair)

external ocular nerve
(VI cranial pair)

facial nerve
(VII cranial pair)

intermediary nerve

auditory or
vestibulocochlear nerve
(VIII cranial pair)

glossopharyngeal nerve
(IX cranial pair)

vagus nerve
(X cranial pair)

hypoglossal nerve
(XII cranial pair)

spinal nerve
(XI cranial pair)

cervical roots

spinal cord

OCCIPITAL POLE

Introduction

The cell

The human body

The locomotive system

The digestive system

The respiratory system

The circulatory system

Blood

Lymph

The nervous system

The senses

The urinary system

The reproductive system

Human reproduction

The endocrine system

The immune system

Alphabetical index

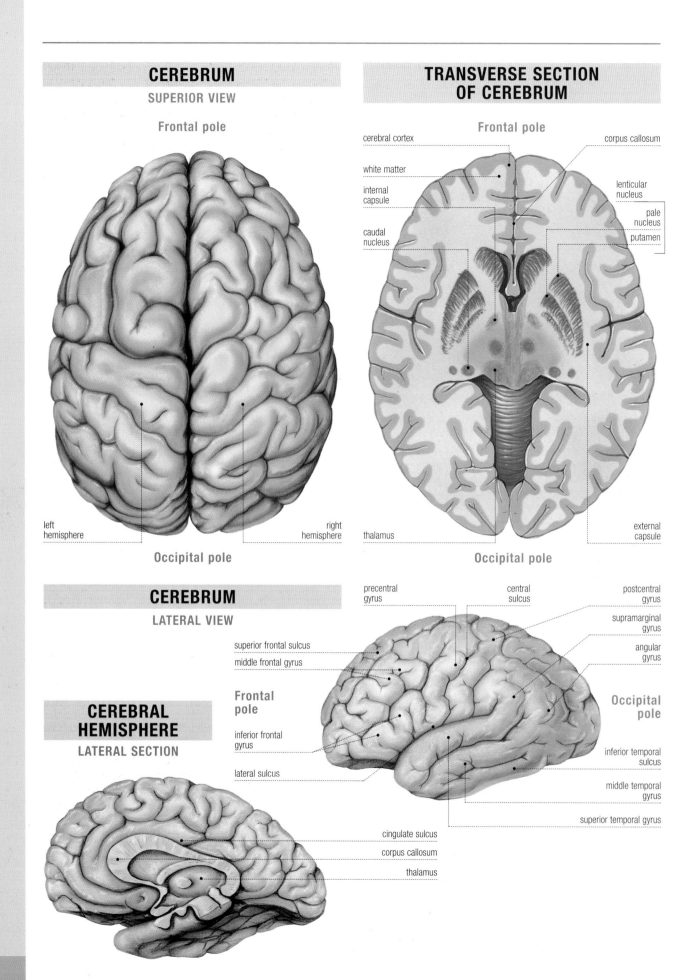

CEREBRUM
SUPERIOR VIEW

Frontal pole

left hemisphere

right hemisphere

Occipital pole

TRANSVERSE SECTION OF CEREBRUM

Frontal pole

cerebral cortex

white matter

internal capsule

caudal nucleus

corpus callosum

lenticular nucleus

pale nucleus

putamen

thalamus

external capsule

Occipital pole

CEREBRUM
LATERAL VIEW

superior frontal sulcus

middle frontal gyrus

Frontal pole

inferior frontal gyrus

lateral sulcus

precentral gyrus

central sulcus

postcentral gyrus

supramarginal gyrus

angular gyrus

Occipital pole

inferior temporal sulcus

middle temporal gyrus

superior temporal gyrus

CEREBRAL HEMISPHERE
LATERAL SECTION

cingulate sulcus

corpus callosum

thalamus

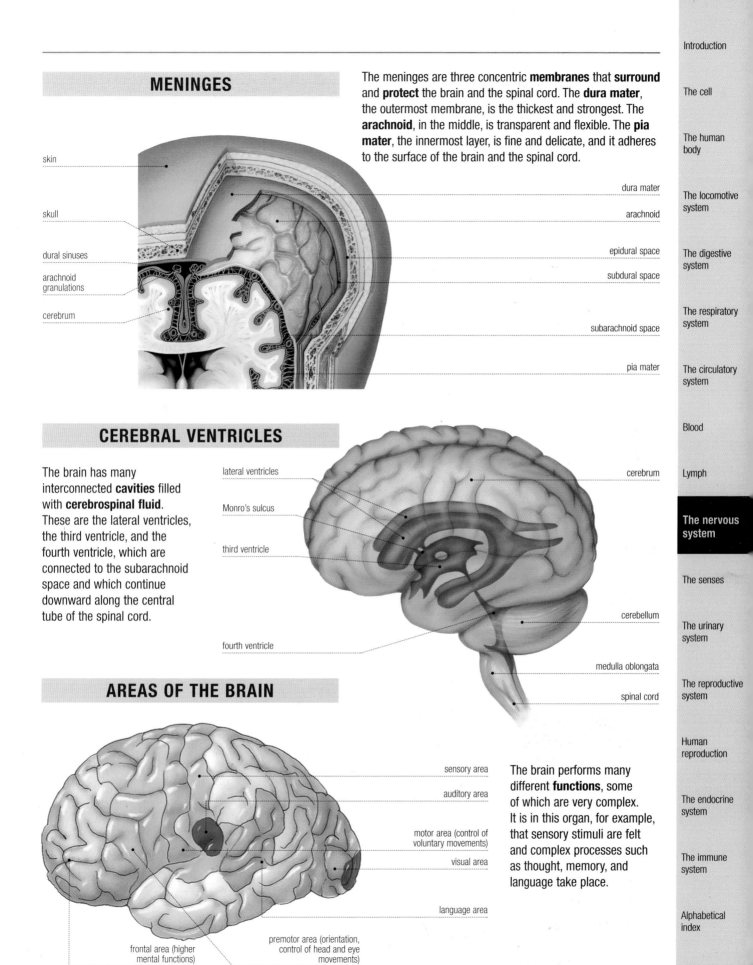

MENINGES

The meninges are three concentric **membranes** that **surround** and **protect** the brain and the spinal cord. The **dura mater**, the outermost membrane, is the thickest and strongest. The **arachnoid**, in the middle, is transparent and flexible. The **pia mater**, the innermost layer, is fine and delicate, and it adheres to the surface of the brain and the spinal cord.

skin

skull

dural sinuses

arachnoid granulations

cerebrum

dura mater

arachnoid

epidural space

subdural space

subarachnoid space

pia mater

CEREBRAL VENTRICLES

The brain has many interconnected **cavities** filled with **cerebrospinal fluid**. These are the lateral ventricles, the third ventricle, and the fourth ventricle, which are connected to the subarachnoid space and which continue downward along the central tube of the spinal cord.

lateral ventricles

Monro's sulcus

third ventricle

fourth ventricle

cerebrum

cerebellum

medulla oblongata

spinal cord

AREAS OF THE BRAIN

sensory area

auditory area

motor area (control of voluntary movements)

visual area

language area

premotor area (orientation, control of head and eye movements)

frontal area (higher mental functions)

The brain performs many different **functions**, some of which are very complex. It is in this organ, for example, that sensory stimuli are felt and complex processes such as thought, memory, and language take place.

Introduction

The cell

The human body

The locomotive system

The digestive system

The respiratory system

The circulatory system

Blood

Lymph

The nervous system

The senses

The urinary system

The reproductive system

Human reproduction

The endocrine system

The immune system

Alphabetical index

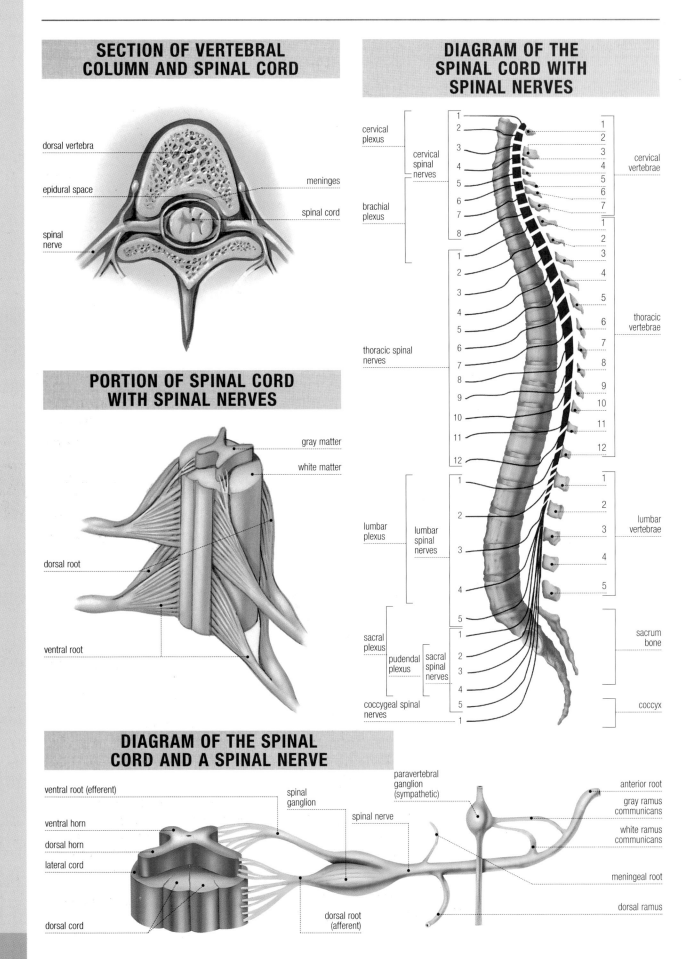

SECTION OF VERTEBRAL COLUMN AND SPINAL CORD

dorsal vertebra

epidural space

spinal nerve

meninges

spinal cord

PORTION OF SPINAL CORD WITH SPINAL NERVES

gray matter

white matter

dorsal root

ventral root

DIAGRAM OF THE SPINAL CORD WITH SPINAL NERVES

cervical plexus

cervical spinal nerves

brachial plexus

thoracic spinal nerves

lumbar plexus

lumbar spinal nerves

sacral plexus

pudendal plexus

sacral spinal nerves

coccygeal spinal nerves

cervical vertebrae

thoracic vertebrae

lumbar vertebrae

sacrum bone

coccyx

DIAGRAM OF THE SPINAL CORD AND A SPINAL NERVE

ventral root (efferent)

ventral horn

dorsal horn

lateral cord

dorsal cord

spinal ganglion

spinal nerve

paravertebral ganglion (sympathetic)

anterior root

gray ramus communicans

white ramus communicans

meningeal root

dorsal ramus

dorsal root (afferent)

PERIPHERAL NERVOUS SYSTEM

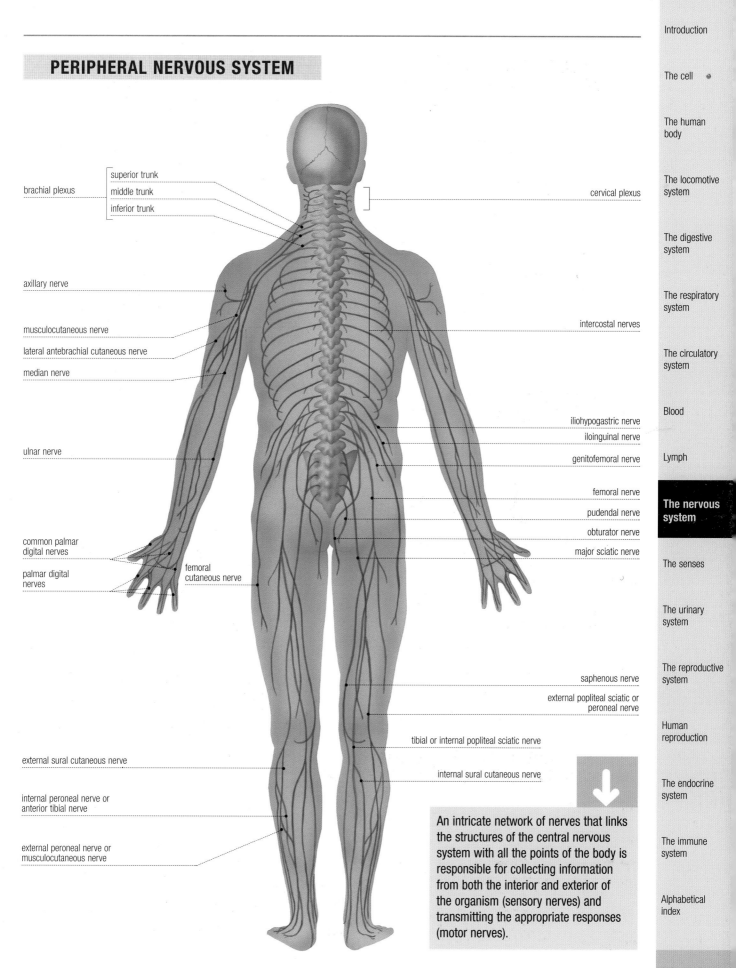

brachial plexus

superior trunk
middle trunk
inferior trunk

cervical plexus

axillary nerve

musculocutaneous nerve

lateral antebrachial cutaneous nerve

median nerve

intercostal nerves

ulnar nerve

iliohypogastric nerve
iloinguinal nerve
genitofemoral nerve

femoral nerve
pudendal nerve
obturator nerve
major sciatic nerve

common palmar digital nerves

palmar digital nerves

femoral cutaneous nerve

saphenous nerve

external popliteal sciatic or peroneal nerve

tibial or internal popliteal sciatic nerve

external sural cutaneous nerve

internal sural cutaneous nerve

internal peroneal nerve or anterior tibial nerve

external peroneal nerve or musculocutaneous nerve

An intricate network of nerves that links the structures of the central nervous system with all the points of the body is responsible for collecting information from both the interior and exterior of the organism (sensory nerves) and transmitting the appropriate responses (motor nerves).

Introduction

The cell

The human body

The locomotive system

The digestive system

The respiratory system

The circulatory system

Blood

Lymph

The nervous system

The senses

The urinary system

The reproductive system

Human reproduction

The endocrine system

The immune system

Alphabetical index

SIGHT

Sight is the sense that gives us the most information about the world around us. The eyes receive **light stimuli** from external sources and transform them into nerve signals that travel along specific routes to the brain, where they are transformed into **visual images**.

LATERAL SECTION OF AN EYE

conjunctiva

transparent membrane that covers the forward part of the sclera and the internal surface of the eyelids

iris

pigmented, muscular disk at the center of which is located an orifice, the pupil; the degree of contraction or dilation of the pupil regulates the passage of rays of light to the back of the eye

cornea

transparent disk through which light rays penetrate to the interior of the eyeball

aqueous humor

transparent liquid that fills the front part of the eye

ciliary body

structure consisting of many muscle fibers that contract to modify the degree of curvature of the lens

lens

transparent, elastic convex body that acts as a lens and focuses light rays onto the retina

sclera

strong, opaque, exterior covering of the eyeball that is visible only at the front (the white of the eye)

choroid

middle layer of the eyeball, containing abundant blood vessels

retina

internal layer of the eyeball, containing light-sensitive cells, and onto which light rays are projected

optic disk

area where the extensions of the cells of the retina that form the optic nerve emerge; it is incapable of vision (blind spot)

macula lutea

small, yellow-colored area of the retina that represents the point of maximum visual clarity

optic nerve

collection of nerve fibers that carry general signals from the retina toward the brain

vitreous humor

gelatinous, transparent mass that takes up most of the interior of the eyeball and allows it to maintain its shape

PROJECTION OF IMAGES ONTO THE RETINA

Light rays from external objects enter the eye by means of the cornea, pass through the pupil, and are focused by the lens onto the retina. There they form an inverted image that is later interpreted by the brain into its original position.

retina

cornea

pupil

lens

projection of image onto the retina

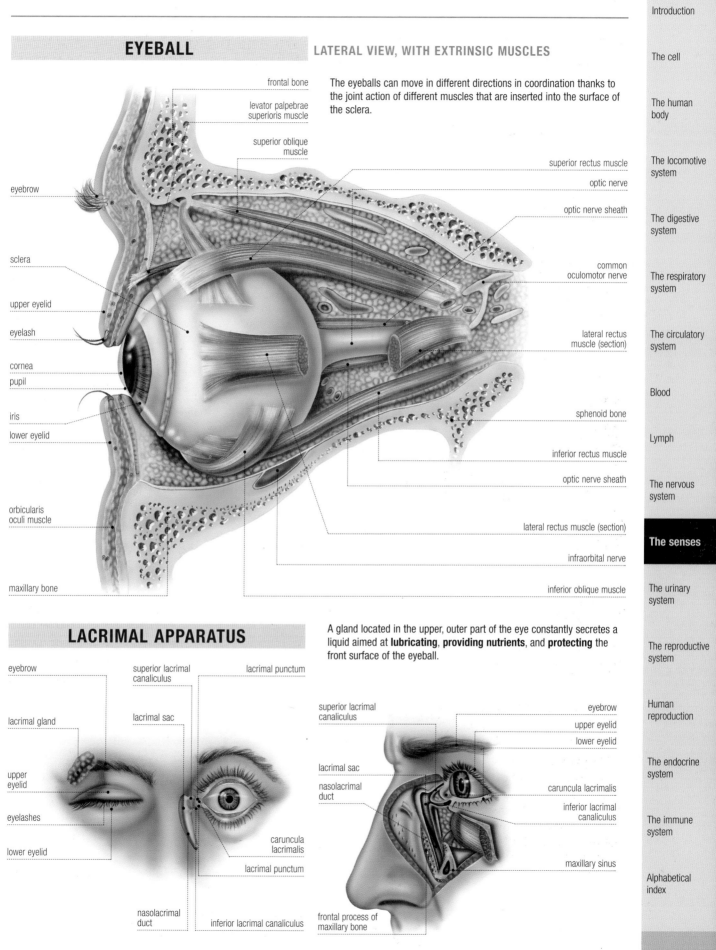

EYEBALL

LATERAL VIEW, WITH EXTRINSIC MUSCLES

The eyeballs can move in different directions in coordination thanks to the joint action of different muscles that are inserted into the surface of the sclera.

frontal bone

levator palpebrae superioris muscle

superior oblique muscle

eyebrow

superior rectus muscle

optic nerve

optic nerve sheath

sclera

common oculomotor nerve

upper eyelid

eyelash

cornea

pupil

lateral rectus muscle (section)

iris

lower eyelid

sphenoid bone

inferior rectus muscle

optic nerve sheath

orbicularis oculi muscle

lateral rectus muscle (section)

infraorbital nerve

maxillary bone

inferior oblique muscle

LACRIMAL APPARATUS

A gland located in the upper, outer part of the eye constantly secretes a liquid aimed at **lubricating**, **providing nutrients**, and **protecting** the front surface of the eyeball.

eyebrow

superior lacrimal canaliculus

lacrimal punctum

lacrimal gland

lacrimal sac

superior lacrimal canaliculus

eyebrow

upper eyelid

lower eyelid

upper eyelid

lacrimal sac

nasolacrimal duct

caruncula lacrimalis

inferior lacrimal canaliculus

eyelashes

lower eyelid

caruncula lacrimalis

lacrimal punctum

maxillary sinus

nasolacrimal duct

inferior lacrimal canaliculus

frontal process of maxillary bone

Introduction

The cell

The human body

The locomotive system

The digestive system

The respiratory system

The circulatory system

Blood

Lymph

The nervous system

The senses

The urinary system

The reproductive system

Human reproduction

The endocrine system

The immune system

Alphabetical index

SIGHT

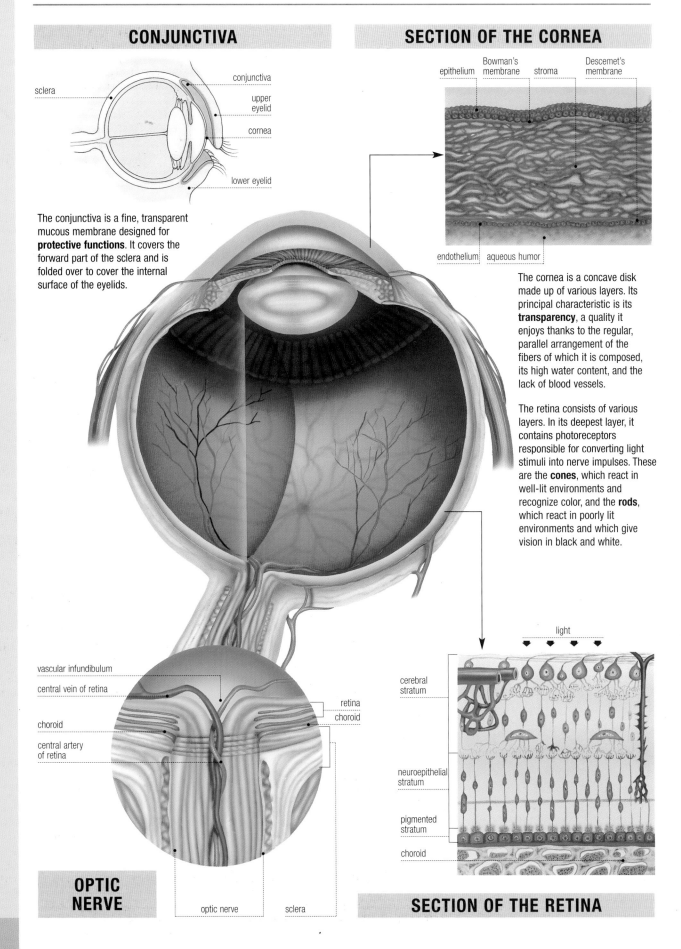

CONJUNCTIVA

sclera

conjunctiva

upper eyelid

cornea

lower eyelid

The conjunctiva is a fine, transparent mucous membrane designed for **protective functions**. It covers the forward part of the sclera and is folded over to cover the internal surface of the eyelids.

SECTION OF THE CORNEA

epithelium | Bowman's membrane | stroma | Descemet's membrane

endothelium | aqueous humor

The cornea is a concave disk made up of various layers. Its principal characteristic is its **transparency**, a quality it enjoys thanks to the regular, parallel arrangement of the fibers of which it is composed, its high water content, and the lack of blood vessels.

The retina consists of various layers. In its deepest layer, it contains photoreceptors responsible for converting light stimuli into nerve impulses. These are the **cones**, which react in well-lit environments and recognize color, and the **rods**, which react in poorly lit environments and which give vision in black and white.

vascular infundibulum

central vein of retina

choroid

central artery of retina

retina
choroid

optic nerve

sclera

OPTIC NERVE

SECTION OF THE RETINA

light

cerebral stratum

neuroepithelial stratum

pigmented stratum

choroid

VISUAL PATHWAYS

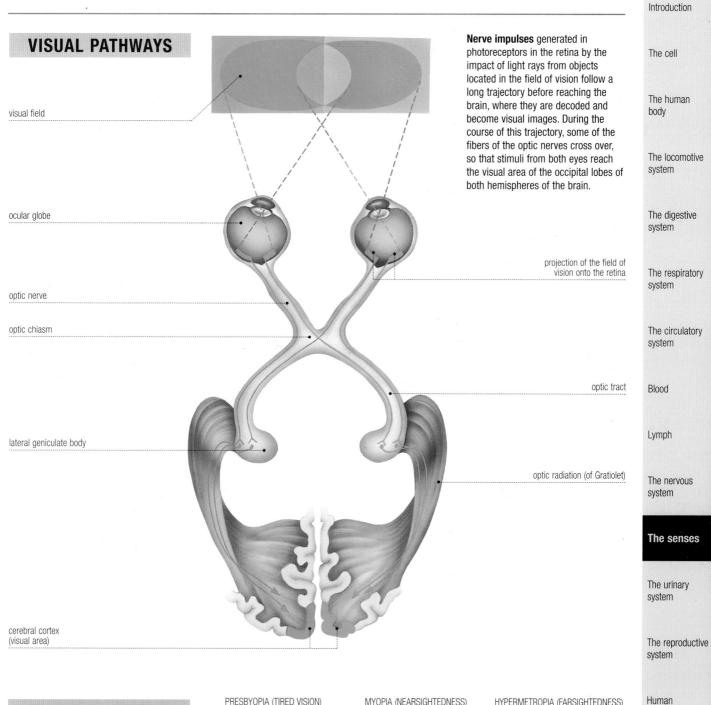

visual field

ocular globe

optic nerve

optic chiasm

lateral geniculate body

cerebral cortex
(visual area)

projection of the field of
vision onto the retina

optic tract

optic radiation (of Gratiolet)

Nerve impulses generated in photoreceptors in the retina by the impact of light rays from objects located in the field of vision follow a long trajectory before reaching the brain, where they are decoded and become visual images. During the course of this trajectory, some of the fibers of the optic nerves cross over, so that stimuli from both eyes reach the visual area of the occipital lobes of both hemispheres of the brain.

PRINCIPAL DEFECTS OF SIGHT AND METHODS OF CORRECTING THEM

PRESBYOPIA (TIRED VISION)

The lens loses elasticity and for this reason does not curve sufficiently. The image of objects nearby forms behind the retina.

A convergent lens compensates for the lack of adaptation of the lens.

MYOPIA (NEARSIGHTEDNESS)

The lens functions well, but the eyeball is too long. The image of distant objects forms in front of the retina.

A divergent lens locates a clear image onto the retina.

HYPERMETROPIA (FARSIGHTEDNESS)

The lens functions well, but the eyeball is too short. The image of nearby objects forms behind the retina.

A convergent lens locates a clear image onto the retina.

Introduction

The cell

The human body

The locomotive system

The digestive system

The respiratory system

The circulatory system

Blood

Lymph

The nervous system

The senses

The urinary system

The reproductive system

Human reproduction

The endocrine system

The immune system

Alphabetical index

HEARING

The ear is a complex organ. It is responsible for hearing, the sense by which we perceive **sounds** coming from our exterior and a fundamental tool for warning us of what is happening in our surroundings, as well as for **communicating** with others. It is also involved in physical **equilibrium**.

There are three sections to the ear:

- **the outer ear** consists of the ear, or pinna, and the auditory canal;

- **the middle ear** is located in a cavity in the temporal bone called the tympanic cavity; it is separated from the outer ear by a vibrating membrane called the tympanic membrane or eardrum; it houses a series of three small bones;

- **the inner ear**, also called the labyrinth, consists of two sections; in the anterior labyrinth, called the cochlea, the hearing organ (organ of Corti) is located; in the posterior labyrinth or vestibule, stimuli that are involved in maintaining physical equilibrium take place.

CROSS SECTION OF THE EAR

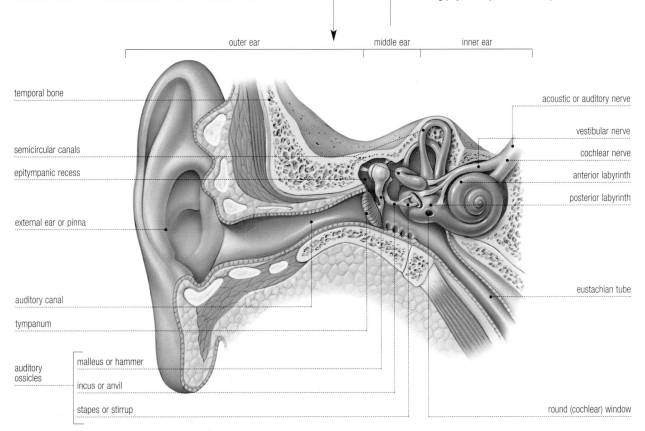

outer ear | middle ear | inner ear

temporal bone
semicircular canals
epitympanic recess
external ear or pinna
auditory canal
tympanum
auditory ossicles
malleus or hammer
incus or anvil
stapes or stirrup

acoustic or auditory nerve
vestibular nerve
cochlear nerve
anterior labyrinth
posterior labyrinth
eustachian tube
round (cochlear) window

EXTERNAL EAR

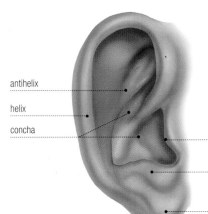

The ear consists of an elastic, skin-covered cartilage. Although it can be found in varying shapes and sizes, it presents a series of characteristic folds and indentations.

antihelix
helix
concha
tragus
antitragus
lobule

OSSICLES OF THE MIDDLE EAR

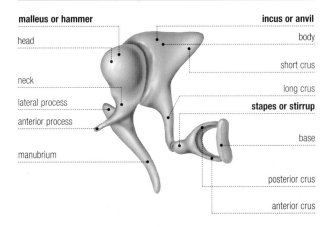

malleus or hammer
head
neck
lateral process
anterior process
manubrium

incus or anvil
body
short crus
long crus
stapes or stirrup
base
posterior crus
anterior crus

Introduction

The cell

The human body

The locomotive system

The digestive system

The respiratory system

The circulatory system

Blood

Lymph

The nervous system

The senses

The urinary system

The reproductive system

Human reproduction

The endocrine system

The immune system

Alphabetical index

THE AUDITORY PROCESS

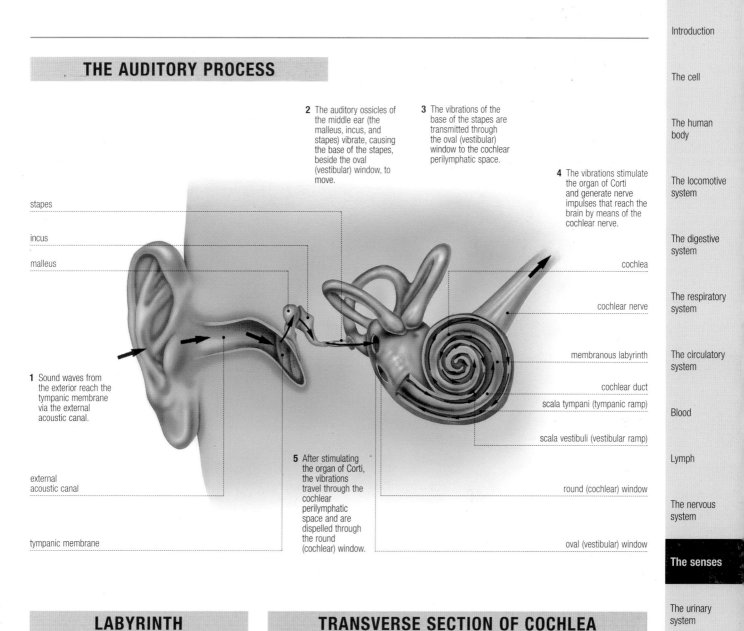

2 The auditory ossicles of the middle ear (the malleus, incus, and stapes) vibrate, causing the base of the stapes, beside the oval (vestibular) window, to move.

3 The vibrations of the base of the stapes are transmitted through the oval (vestibular) window to the cochlear perilymphatic space.

4 The vibrations stimulate the organ of Corti and generate nerve impulses that reach the brain by means of the cochlear nerve.

stapes

incus

malleus

1 Sound waves from the exterior reach the tympanic membrane via the external acoustic canal.

external acoustic canal

tympanic membrane

5 After stimulating the organ of Corti, the vibrations travel through the cochlear perilymphatic space and are dispelled through the round (cochlear) window.

cochlea

cochlear nerve

membranous labyrinth

cochlear duct

scala tympani (tympanic ramp)

scala vestibuli (vestibular ramp)

round (cochlear) window

oval (vestibular) window

LABYRINTH

anterior semicircular canal

vestibule

posterior semicircular canal

lateral semicircular canal

cochlea

oval (vestibular) window

round (cochlear) window

TRANSVERSE SECTION OF COCHLEA

scala vestibuli (vestibular ramp)

cochlear duct

tectorial membrane

vestibular ramp

Reissner's membrane

cochlear duct

organ of Corti

basilar membrane

tympanic membrane

scala tympani (tympanic ramp)

cochlear nerve

63

SMELL

Smell is a sense that has various functions. It is involved in the digestive process, because the pleasant **smell** of food stimulates salivary and gastric secretions. It warns us of the presence of toxic gases that smell bad. It provides **sensations**, either pleasurable or disagreeable, that affect our emotional life.

ethmoid bone
olfactory bulb

olfactory nerve

olfactory cells
olfactory gland

support cells

LOCATION OF THE OLFACTORY MEMBRANE AND THE OLFACTORY CELL

In the mucous membrane that covers the ceiling of the nasal passages is an area about 0.4 square inch (2.5 cm²) called the olfactory membrane. It is covered with numerous specialized cells that detect **volatile olfactory substances** in the air we breathe.

olfactory bulb
ethmoid bone
olfactory nerve

The olfactory cells are interspersed among support cells and small, mucus-producing glands. Each one has a free end with several minute **cilia** that react to contact with olfactory substances and generate **impulses**. These impulses travel along a narrow nerve fiber that emerges from the other end and passes through the ethmoid bones before reaching the olfactory bulb.

OLFACTORY CELL

nerve fibers

nucleus

olfactory cilia

SECTION OF THE OLFACTORY MEMBRANE

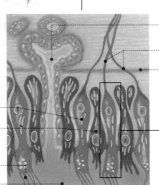

olfactory gland
nerve fibers
basal membrane

olfactory cells
support cells
olfactory cilia
mucus

TASTE

Introduction

The cell

The human body

The locomotive system

The digestive system

The respiratory system

The circulatory system

Blood

Lymph

The nervous system

The senses

The urinary system

The reproductive system

Human reproduction

The endocrine system

The immune system

Alphabetical index

THE TONGUE
DORSAL VIEW

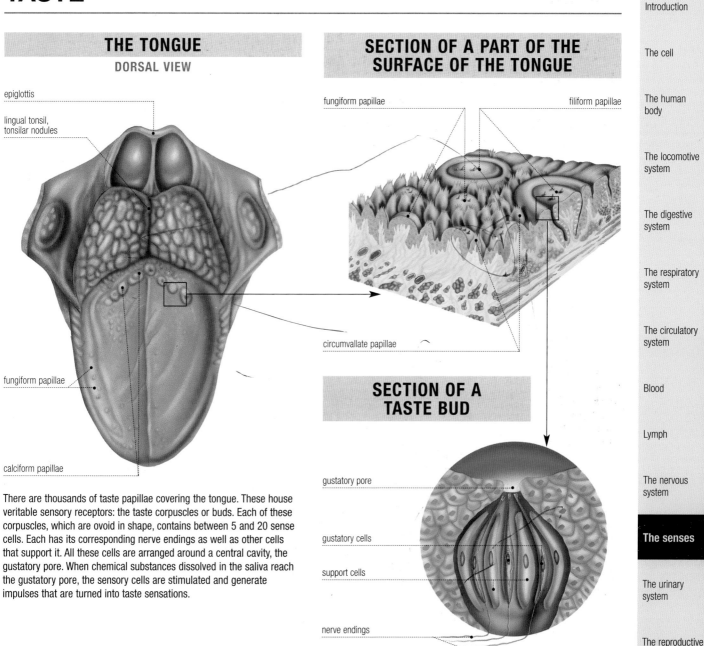

epiglottis

lingual tonsil, tonsilar nodules

fungiform papillae

calciform papillae

SECTION OF A PART OF THE SURFACE OF THE TONGUE

fungiform papillae

filiform papillae

circumvallate papillae

There are thousands of taste papillae covering the tongue. These house veritable sensory receptors: the taste corpuscles or buds. Each of these corpuscles, which are ovoid in shape, contains between 5 and 20 sense cells. Each has its corresponding nerve endings as well as other cells that support it. All these cells are arranged around a central cavity, the gustatory pore. When chemical substances dissolved in the saliva reach the gustatory pore, the sensory cells are stimulated and generate impulses that are turned into taste sensations.

SECTION OF A TASTE BUD

gustatory pore

gustatory cells

support cells

nerve endings

AREAS OF PERCEPTION OF DIFFERENT TASTES

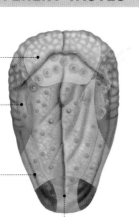

(yellow)
area of bitter perception

(blue)
area of acid or sour perception

(green)
area of salt perception

(red)
area of sweet perception

TYPES OF GUSTATORY PAPILLAE

There are different types of gustatory papillae, though all of them perceive the basic sensations: **sweet**, **salt**, **acid** and **sour**, and **bitter**. However, the different papillae, distributed unevenly over the surface of the tongue, respond with greater or lesser intensity to the different stimuli. For this reason, some areas of the tongue perceive a particular taste better than others.

fungiform papilla

circumvallate papilla

filiform papilla

conical papilla

lenticular papilla

TOUCH

The skin represents the **covering** of our bodies. It is fully equipped with sensory nerves capable of detecting a wide range of external **stimuli** and providing us with important **information** concerning the environment around us. For this reason, it has plenty of receptors responsible for immediately detecting tactile, thermal (hot and cold), or painful stimuli.

SECTION OF THE SKIN

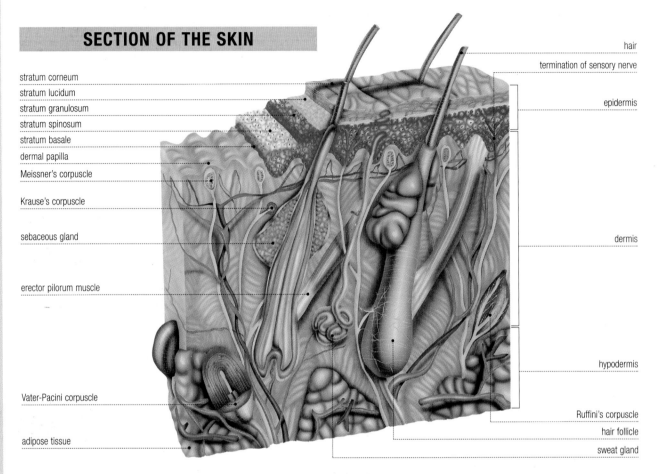

stratum corneum
stratum lucidum
stratum granulosum
stratum spinosum
stratum basale
dermal papilla
Meissner's corpuscle
Krause's corpuscle
sebaceous gland
erector pilorum muscle
Vater-Pacini corpuscle
adipose tissue

hair
termination of sensory nerve
epidermis
dermis
hypodermis
Ruffini's corpuscle
hair follicle
sweat gland

The skin is a strong, flexible **membrane** that covers the whole body and **protects** the organism from aggressive external agents. It is involved in such important functions as **regulating temperature control** and acting as a veritable **sensory organ**. It consists of three superimposed layers:

• the **epidermis**, the outermost layer, is in direct contact with the exterior;

• the **dermis**, lying underneath the epidermis, is essentially made up of connective tissue;

• the **hypodermis**, or subcutaneous cellular tissue, is the deepest layer and primarily consists of adipose tissue (fat), which isolates the body from cold, softens any impact, and acts as the body's main energy reserve.

SENSORY RECEPTORS

touch receptor
(Meissner's corpuscle)

receptor for pressure
and vibration
(Vater-Pacini corpuscle)

receptor for heat
(Ruffini's corpuscle)

receptor for cold
(Krause's corpuscle)

receptor for pain
(free nerve ending)

A multitude of sensory receptors are distributed over the entire surface of the skin, although different parts of the body have varying concentrations. They respond to various stimuli and send the relevant information to the nervous system to be suitably interpreted.

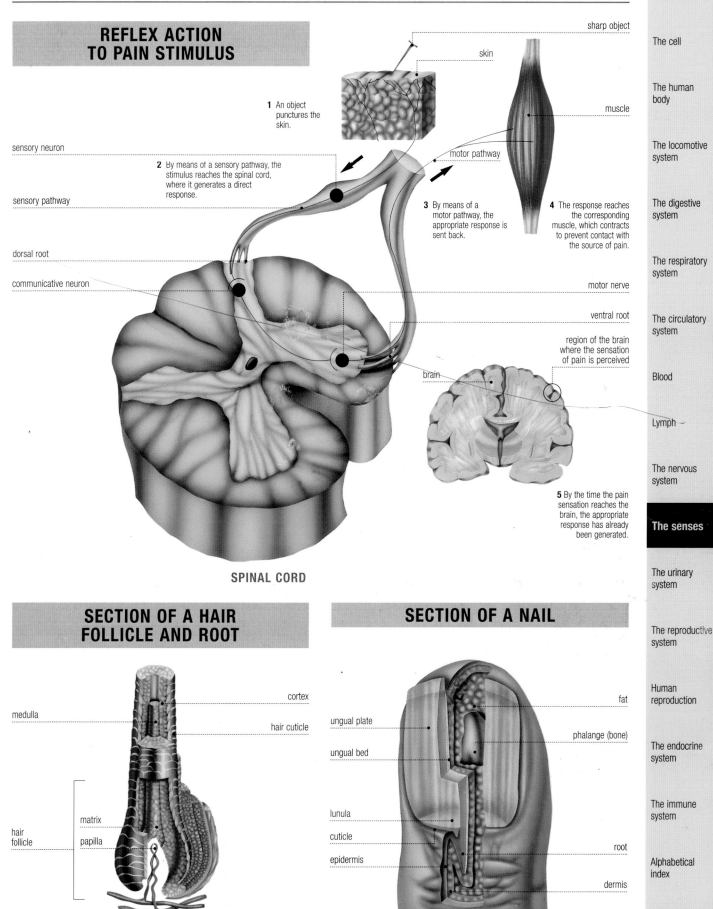

REFLEX ACTION TO PAIN STIMULUS

sharp object

skin

1 An object punctures the skin.

muscle

sensory neuron

motor pathway

2 By means of a sensory pathway, the stimulus reaches the spinal cord, where it generates a direct response.

sensory pathway

3 By means of a motor pathway, the appropriate response is sent back.

4 The response reaches the corresponding muscle, which contracts to prevent contact with the source of pain.

dorsal root

communicative neuron

motor nerve

ventral root

region of the brain where the sensation of pain is perceived

brain

5 By the time the pain sensation reaches the brain, the appropriate response has already been generated.

SPINAL CORD

SECTION OF A HAIR FOLLICLE AND ROOT

cortex

medulla

hair cuticle

matrix

hair follicle

papilla

SECTION OF A NAIL

fat

ungual plate

phalange (bone)

ungual bed

lunula

cuticle

epidermis

root

dermis

Introduction

The cell

The human body

The locomotive system

The digestive system

The respiratory system

The circulatory system

Blood

Lymph

The nervous system

The senses

The urinary system

The reproductive system

Human reproduction

The endocrine system

The immune system

Alphabetical index

THE URINARY SYSTEM

The urinary system is made up of various organs whose purpose is to **filter the blood** in order to regulate its composition and purify it of toxic waste products. At the same time, it eliminates excess water and gets rid of toxic residues from the body by means of urine.

COMPONENTS OF THE URINARY APPARATUS

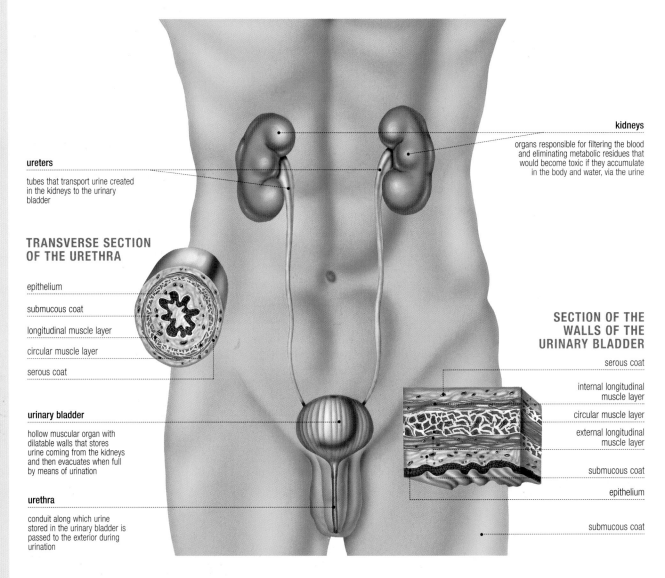

kidneys
organs responsible for filtering the blood and eliminating metabolic residues that would become toxic if they accumulate in the body and water, via the urine

ureters
tubes that transport urine created in the kidneys to the urinary bladder

TRANSVERSE SECTION OF THE URETHRA

epithelium

submucous coat

longitudinal muscle layer

circular muscle layer

serous coat

SECTION OF THE WALLS OF THE URINARY BLADDER

serous coat

internal longitudinal muscle layer

circular muscle layer

external longitudinal muscle layer

submucous coat

epithelium

submucous coat

urinary bladder
hollow muscular organ with dilatable walls that stores urine coming from the kidneys and then evacuates when full by means of urination

urethra
conduit along which urine stored in the urinary bladder is passed to the exterior during urination

RENAL CIRCULATION

Blood circulating in the organism passes repeatedly through the kidneys. These organs must **eliminate** the **toxic residues** that are constantly produced by cellular metabolism, via the urine. For this reason, the amount of blood that reaches the kidneys in a given period of time is very high. It represents around 20 percent of the total volume propelled by the heart. Each minute, around 1.3 quarts (1.2 l) of blood circulate through the kidneys.

lung

pulmonary artery

pulmonary vein

heart

aortic artery

inferior vena cava

kidney

ANTERIOR VIEW OF THE KIDNEYS WITH BLOOD VESSELS

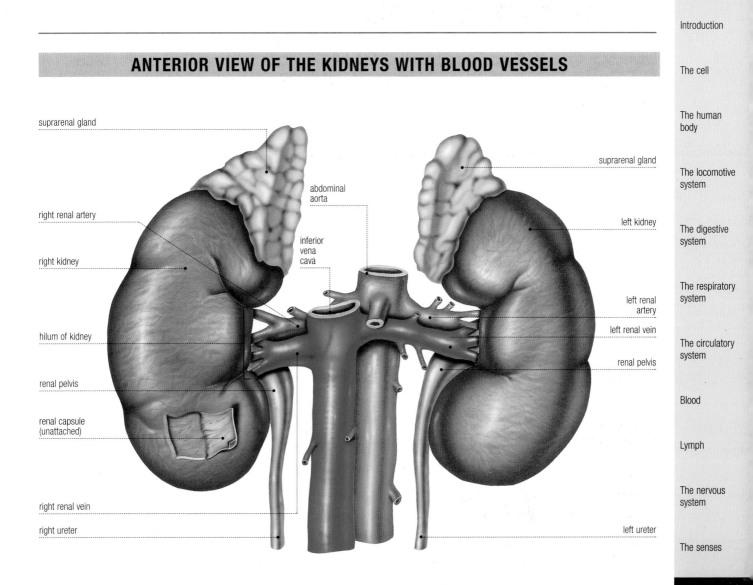

suprarenal gland

right renal artery

right kidney

hilum of kidney

renal pelvis

renal capsule
(unattached)

right renal vein

right ureter

abdominal
aorta

inferior
vena
cava

suprarenal gland

left kidney

left renal
artery

left renal vein

renal pelvis

left ureter

Introduction

The cell

The human
body

The locomotive
system

The digestive
system

The respiratory
system

The circulatory
system

Blood

Lymph

The nervous
system

The senses

**The urinary
system**

The
reproductive
system

Human
reproduction

The endocrine
system

The immune
system

Alphabetical
index

SECTION OF KIDNEY

(LEFT)

The familiar bean shape of the kidney is covered in a fibrous capsule. Inside, two distinct areas can be distinguished: a peripheral, yellow-colored area, the **renal cortex**, and a dark red, inner area, the **renal medulla**. Here are found between 12 and 15 triangular, cone-shaped structures, called the **pyramids of Malpighi**. These are separated by extensions of the cortex that extend down into the medulla and are called the **columns of Bertin**. The bases of the pyramids point toward the outside of the kidney, while the tips point into the center, which is hollow and is known as the renal sinus. At the point of each pyramid, or **papilla**, are two minute orifices through which the urine made in the kidney passes into fine tubes, called **minor calyxes**, that then empty into larger tubes, called **major calyxes.** These flow together to form a cavity in the form of a funnel, called the **renal pelvis**, that exits via the internal edge of the kidney and continues as the ureter.

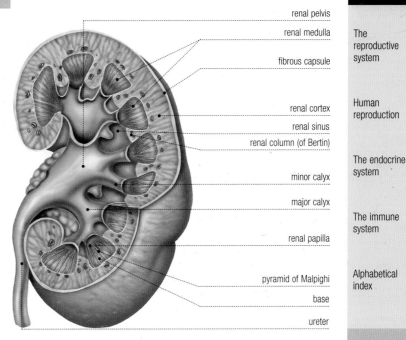

renal pelvis

renal medulla

fibrous capsule

renal cortex

renal sinus

renal column (of Bertin)

minor calyx

major calyx

renal papilla

pyramid of Malpighi

base

ureter

BLOOD VESSELS OF THE KIDNEY

Within the kidney, the renal artery subdivides several times so that only a small arteriole reaches each of the functional units of the organ, the nephrons. Each of these consists of two parts: a corpuscle where **blood is filtered** and a tubule where **urine is produced**. An afferent arteriole reaches each corpuscle and divides into numerous capillaries that make up a cluster, called the glomerulus, surrounded by a double membrane in the shape of a funnel, called the Bowman's capsule. Blood circulates through the capillaries of the glomerule, which has tiny pores in its walls that filter liquid and small molecules. The Bowman's capsule collects the filtrate and empties it into the renal tubule, a tube with different sections along which most of the water and some useful substances are reabsorbed while other harmful substances that have not already been filtered out are eliminated, and urine is created.

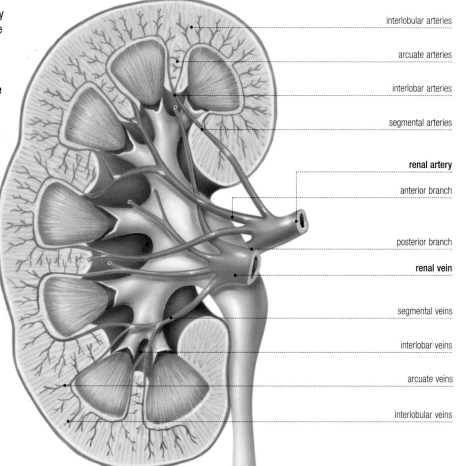

interlobular arteries

arcuate arteries

interlobar arteries

segmental arteries

renal artery

anterior branch

posterior branch

renal vein

segmental veins

interlobar veins

arcuate veins

interlobular veins

DIAGRAM OF A NEPHRON

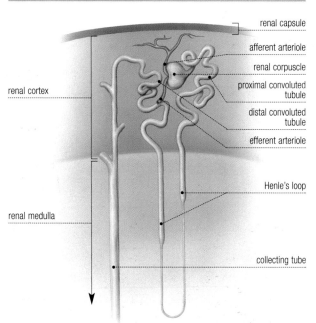

renal capsule

afferent arteriole

renal corpuscle

proximal convoluted tubule

distal convoluted tubule

efferent arteriole

renal cortex

Henle's loop

renal medulla

collecting tube

The kidneys have an extraordinary working capacity. If they are diseased, only 25 to 30 percent of the nephrons need to remain unaffected to ensure the correct production of urine.

DIAGRAM OF A NEPHRON

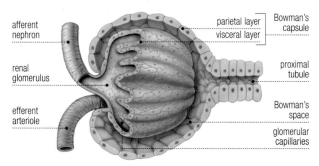

afferent nephron

renal glomerulus

efferent arteriole

parietal layer
visceral layer

Bowman's capsule

proximal tubule

Bowman's space

glomerular capillaries

URINARY BLADDER
POSTERIOR VIEW

- right ureter
- left ureter
- muscular sheath
- vas deferens
- seminal vesicle
- prostate gland

SECTION OF THE MASCULINE URINARY BLADDER
ANTERIOR VIEW

- bladder wall
- peritoneum
- muscular sheath
- body of urinary bladder
- mucous membrane
- opening of urethra
- trigone
- urethral crest
- internal opening of urethra
- prostate

FULL AND EMPTY URINARY BLADDER

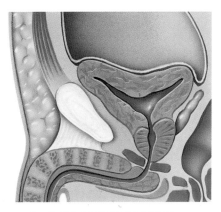

EMPTY URINARY BLADDER

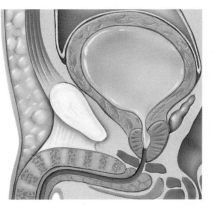

FULL URINARY BLADDER

The urinary bladder is located in the center of the pelvic cavity and is similar in both sexes, although the relationship with adjacent organs is different in men and women. When the urinary bladder is empty, it is triangular in shape. As it fills up, it adopts an oval or spherical shape. In adults, the urinary bladder can store up to 12 fluid ounces (350 ml) of urine.

MALE URETHRA
LATERAL VIEW

- urinary bladder
- prostate
- prostatic section of urethra
- membranous section of urethra
- spongy section of urethra
- corpus cavernosum
- glans
- seminal vesicle
- seminal duct
- bulbourethral (Cowper's) gland
- urethral sphincter
- corpus spongiosum
- navicular fossa

FEMALE URETHRA
ANTERIOR VIEW

- urinary bladder
- urethra
- orifices of periurethral glands
- urethral sphincter
- vagina
- labia minora
- labia majora

Introduction

The cell

The human body

The locomotive system

The digestive system

The respiratory system

The circulatory system

Blood

Lymph

The nervous system

The senses

The urinary system

The reproductive system

Human reproduction

The endocrine system

The immune system

Alphabetical index

THE MALE REPRODUCTIVE SYSTEM

The male reproductive system consists of a series of genital organs, some internal, some external, that allow the man to take part in the process of **pro-creation** and are perfectly adapted for effectively taking part in **sexual activity**.

MALE GENITAL ORGANS

abdominal cavity

peritoneum

urinary bladder

ductus deferens

a tube responsible for transporting sperm from the epididymis toward the exterior

pubic symphysis

prostate

gland in the form of a chestnut, responsible for secreting a component of semen consisting of nutritive elements for the spermatozoa

epididymis

tubular structure located around the testicle, where sperm are fabricated and matured

urethra

tube along which semen is expelled at the moment of ejaculation

penis

cylindrical organ equipped with erectile bodies that allow it to increase in size and stiffness, which is vital for copulation to take place.

corpus cavernosum

corpus spongiosum

glans

reservoir of ductus deferens

dilation of the final part of the ductus deferens where sperm are stored

seminal vesicle

small tubular gland that produces a yellowish, viscous secretion that is one of the components of seminal fluid

rectum

ejaculatory duct

a tube that receives the sperm from the vas deferens as well as secretions from the seminal vesicle and that passes through the prostate gland before opening into the urethra

bulbourethral (Cowper's) gland

anus

testicle

glandular, oval-shaped organ of the male gonad, where sperm and the sexual hormone testosterone are produced

scrotum

cutaneous sac, located outside the abdominal cavity, containing the testicles

EXTERNAL MASCULINE GENITALIA

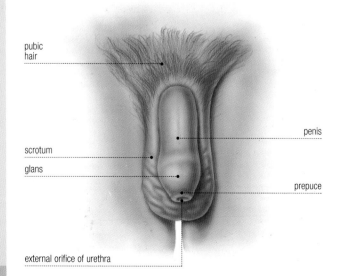

pubic hair

penis

scrotum

glans

prepuce

external orifice of urethra

LOCATION OF THE MALE GENITAL ORGANS

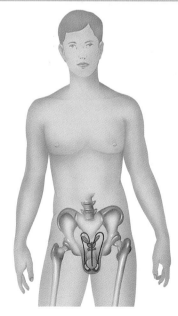

PENIS

The penis is an organ that can increase notably in size and stiffness when erect, thanks to the presence of cylindrical bodies within it that, under certain stimuli, fill with blood. These are the two **corpora cavernosa**, which are symmetrical and are located next to one another in the upper part of the penis, and the **corpus spongiosum**, which is located in the center and beneath the corpora cavernosa and through which the urethra runs lengthways.

LONGITUDINAL SECTION

urinary bladder

prostate

bulbourethral (Cowper's) gland

urethra

corpus cavernosum

corpus cavernosum

corpus spongiosum

glans

prepuce

external orifice of urethra

TRANSVERSE SECTIONS

glans

skin

skin

urethra

corpus cavernosum

urethra

corpus cavernosum

corpus spongiosum

corpus spongiosum

TESTICLE AND EPIDIDYMIS

SECTION

Each testicle is covered in a fibrous membrane called the **tunica albuginea**. Inside, each testicle is divided into various lobules, separated by walls, or septa, of connective tissue that enclose a variable number of **seminiferous tubules**. These are short tubes where sperm is produced. They run together to form a network of ducts that lead into wider ducts, called the **efferent ducts**, that, in turn, empty into the epididymis.

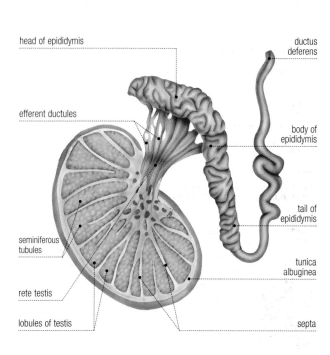

head of epididymis

ductus deferens

efferent ductules

body of epididymis

seminiferous tubules

tail of epididymis

tunica albuginea

rete testis

lobules of testis

septa

PROSTATE

The prostate is a gland located beneath the urinary bladder. The urethra passes through its center, and the ejaculatory ducts leading into the urethra pass through its posterior part. It consists of a multitude of tubular structures whose walls produce a **secretion** that is one of the components of **semen**. The various tubules flow together to form about 20 ducts that empty into the urethra through individual openings. In the moment prior to ejaculation, they empty prostatic secretions into the urethra. At the same time, the ejaculatory ducts empty the liquid from the seminal vesicles and sperm from the testicles into the urethra as well.

ANTERIOR SECTION

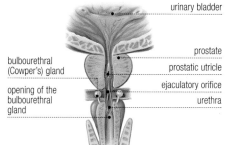

urinary bladder

prostate

bulbourethral (Cowper's) gland

prostatic utricle

opening of the bulbourethral gland

ejaculatory orifice

urethra

LATERAL SECTION

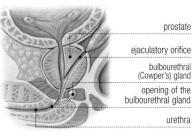

prostate

ejaculatory orifice

bulbourethral (Cowper's) gland

opening of the bulbourethral gland

urethra

TRANSVERSE SECTION

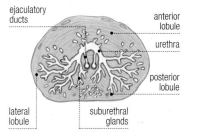

ejaculatory ducts

anterior lobule

urethra

posterior lobule

lateral lobule

suburethral glands

Introduction

The cell

The human body

The locomotive system

The digestive system

The respiratory system

The circulatory system

Blood

Lymph

The nervous system

The senses

The urinary system

The reproductive system

Human reproduction

The endocrine system

The immune system

Alphabetical index

73

THE FEMALE REPRODUCTIVE SYSTEM

The female reproductive system consists of a series of genital organs that allow the woman to take part in the process of **procreation** and are perfectly adapted for effectively taking part in **sexual activity**.

In addition, they include the breasts, the glands that are responsible for producing milk, which is the perfect form of nourishment for the newborn baby.

FEMALE GENITAL ORGANS

uterus

a hollow organ in the shape of an inverted pear; it has thick walls (myometrium) whose internal surface is covered in a layer of mucous membrane (endometrium), which in each menstrual cycle thickens and then is shed as part of menstruation; its purpose is to receive the fertilized ovum and to protect the fetus throughout pregnancy

Fallopian tube

tube with the form of a horn, with the narrow end opening out into the uterus (isthmus) and the funnel-shaped end opening out above the ovary; its purpose is to receive the ovum that is released during ovulation and transport it toward the uterine cavity

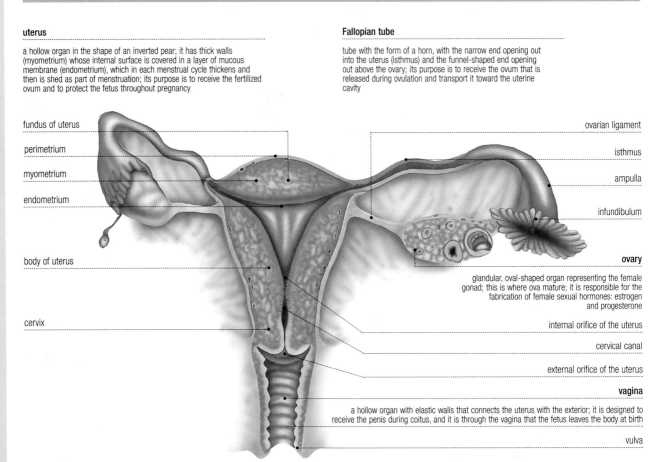

fundus of uterus

perimetrium

myometrium

endometrium

body of uterus

cervix

ovarian ligament

isthmus

ampulla

infundibulum

ovary

glandular, oval-shaped organ representing the female gonad; this is where ova mature; it is responsible for the fabrication of female sexual hormones: estrogen and progesterone

internal orifice of the uterus

cervical canal

external orifice of the uterus

vagina

a hollow organ with elastic walls that connects the uterus with the exterior; it is designed to receive the penis during coitus, and it is through the vagina that the fetus leaves the body at birth

vulva

FEMALE EXTERNAL GENITALIA

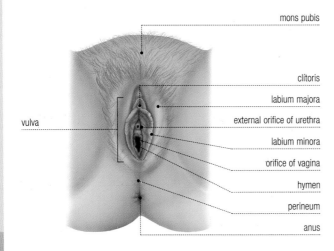

mons pubis

clítoris

labium majora

external orifice of urethra

labium minora

orifice of vagina

hymen

perineum

anus

vulva

LOCATION OF THE FEMALE GENITAL ORGANS

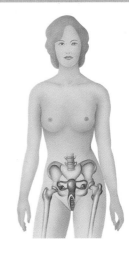

FEMALE GENITAL SYSTEM

LATERAL SECTION

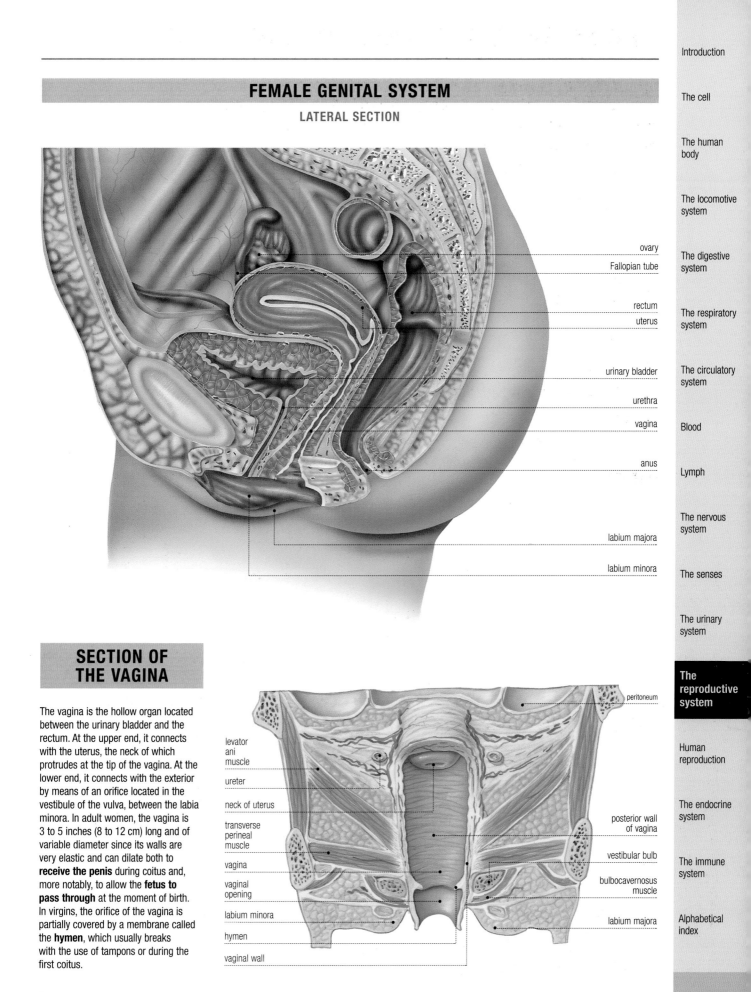

ovary

Fallopian tube

rectum

uterus

urinary bladder

urethra

vagina

anus

labium majora

labium minora

Introduction

The cell

The human body

The locomotive system

The digestive system

The respiratory system

The circulatory system

Blood

Lymph

The nervous system

The senses

The urinary system

The reproductive system

Human reproduction

The endocrine system

The immune system

Alphabetical index

SECTION OF THE VAGINA

The vagina is the hollow organ located between the urinary bladder and the rectum. At the upper end, it connects with the uterus, the neck of which protrudes at the tip of the vagina. At the lower end, it connects with the exterior by means of an orifice located in the vestibule of the vulva, between the labia minora. In adult women, the vagina is 3 to 5 inches (8 to 12 cm) long and of variable diameter since its walls are very elastic and can dilate both to **receive the penis** during coitus and, more notably, to allow the **fetus to pass through** at the moment of birth. In virgins, the orifice of the vagina is partially covered by a membrane called the **hymen**, which usually breaks with the use of tampons or during the first coitus.

peritoneum

levator ani muscle

ureter

neck of uterus

transverse perineal muscle

vagina

vaginal opening

labium minora

hymen

vaginal wall

posterior wall of vagina

vestibular bulb

bulbocavernosus muscle

labium majora

SECTION OF AN OVARY AND EVOLUTION OF AN OVARIAN FOLLICLE

At the moment of birth, the ovary contains thousands of **primary follicles** that store immature female reproductive cells, or oocytes. From puberty onward, various primary follicles capable of secreting estrogen are developed in cycles. At the same time, the primary egg cells they contain within them begin to mature. About 14 days after the beginning of the cycle, one of the follicles completes its development and **ovulation** takes place. The follicle ruptures, and the mature oocyte, now converted into an ovum, is freed from the ovary. The walls of the ruptured follicle then transform into the corpus luteum, which also secretes progesterone. If fertilization does not take place, the **corpus luteum** atrophies. After about 10 to 14 days, it transforms into a white body, the corpus albicans, that stops producing female hormones.

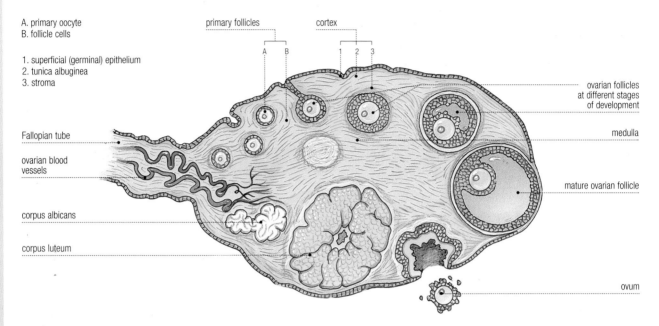

A. primary oocyte
B. follicle cells

1. superficial (germinal) epithelium
2. tunica albuginea
3. stroma

primary follicles

cortex

Fallopian tube

ovarian blood vessels

corpus albicans

corpus luteum

ovarian follicles at different stages of development

medulla

mature ovarian follicle

ovum

BREASTS

ANTERIOR VIEW

submammary fold

Montgomery's tubercles

areola

nipple

During puberty, the female breasts enlarge. **Mammary glands** within them develop in order to **produce milk** to nourish the newborn child in the event of pregnancy. Each mammary gland consists of numerous **acini**, small sacs covered in cells that, under the right hormonal influence, are able to fabricate maternal milk. These **acini** are contained within fatty tissue and empty out into narrow channels that flow into larger ones, called **lactiferous ducts**, that flow in the direction of the exterior. After passing through dilations called **lactiferous sinuses**, they open out at the nipple.

BREAST

LATERAL SECTION

fatty tissue

lactiferous ducts

nipple

lactiferous sinuses

acini of the mammary gland

MENSTRUAL CYCLE

The menstrual cycle is a period that lasts from the first day of one menstruation to the first day of the next, a total of around **28 days**. During the first part, or proliferative phase, of the cycle, estrogen produced by the ovarian follicles makes the mucous membrane that covers the uterus (endometrium) grow thicker and more vascular. This phase lasts until ovulation takes place, around the fourteenth day. In the second part of the cycle, the secretory phase, the progesterone produced by the corpus luteum makes the endometrium increase in thickness in preparation for eventually receiving a fertilized ovum. If fertilization does not take place, the production of female hormones stops. As a consequence, the endometrium is discharged, causing the start of menstrual bleeding that marks the beginning of the next cycle.

VARIABLE DURATION

The menstrual cycle is repeated continuously from puberty to menopause, except during pregnancy. It lasts an average of 28 days, although anything from 21 to 35 days is considered completely normal.

PHASES OF THE CYCLE

| 1 | 2 | 3 | 4 | 5 | 6 | 7 | 8 | 9 | 10 | 11 | 12 | 13 | 14 | 15 | 16 | 17 | 18 | 19 | 20 | 21 | 22 | 23 | 24 | 25 | 26 | 27 | 28 | 1 | 2 | 3 | 4 |

FERTILE PERIOD

1. Menstruation, of variable duration.

2. Phase of growth of the follicle (can last more than 14 days in long cycles).

3. Ovulation, the timing of which is impossible to predict beforehand.

4. Phase of secretion of the yellow body, which only the temperature curve can determine precisely.

1. Menstruation, of variable duration.

THE OVARY

primary oocyte

developing follicle

mature follicle

expulsion of the ovum

corpus luteum

corpus albicans

primary oocyte

THE UTERUS

proliferation of the endometrium

ovulation

discharge of the endometrium

BASAL TEMPERATURE (in degrees Celsius)

| 37,3° |
| 37,2° |
| 37,1° |
| 37° |
| 36,9° |
| 36,8° |
| 36,7° |

| 1 | 2 | 3 | 4 | 5 | 6 | 7 | 8 | 9 | 10 | 11 | 12 | 13 | 14 | 15 | 16 | 17 | 18 | 19 | 20 | 21 | 22 | 23 | 24 | 25 | 26 | 27 | 28 | 1 | 2 | 3 | 4 |

— menstruation —   —proliferative phase—   —secretory phase—   — menstruation —

Introduction

The cell

The human body

The locomotive system

The digestive system

The respiratory system

The circulatory system

Blood

Lymph

The nervous system

The senses

The urinary system

The reproductive system

Human reproduction

The endocrine system

The immune system

Alphabetical index

FERTILIZATION

Fertilization is the union of **seed cells** from both sexes. In other words, an **ovum** from the mother and a **spermatozoon**, or sperm cell, from the father fuse together to make a fertilized egg cell or zygote, from which the new being will grow.

MAN AND WOMAN IN COITUS

During coitus, the man ejaculates, depositing millions of sperm cells into the woman's vagina, which then begin a long journey through the female genital system. If coitus takes place during the woman's fertile period and the sperm cells come across an ovum, one of them will very probably fertilize it.

SPERMATOZOON

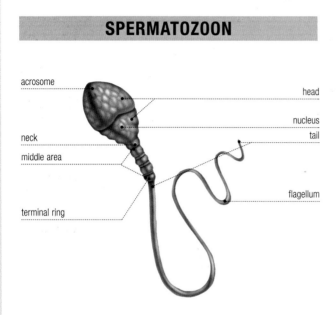

acrosome
head
nucleus
neck
tail
middle area
flagellum
terminal ring

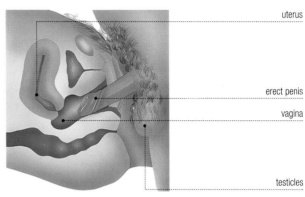

uterus
erect penis
vagina
testicles

OVUM

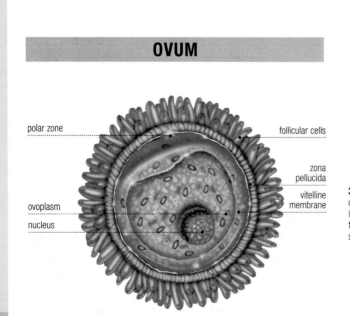

polar zone
follicular cells
zona pellucida
vitelline membrane
ovoplasm
nucleus

UNION OF AN OVUM AND A SPERMATOZOON

1 The head of the spermatozoon presses against the membrane of the ovum

2 The spermatozoon begins to enter the interior of the ovum

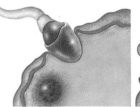

3 The membrane of the ovum repairs itself to prevent fertilization by another spermatozoon

4 The tail of the spermatozoon becomes detached, and only the head penetrates the ovum

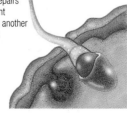

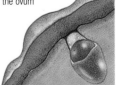

HUMAN REPRODUCTION

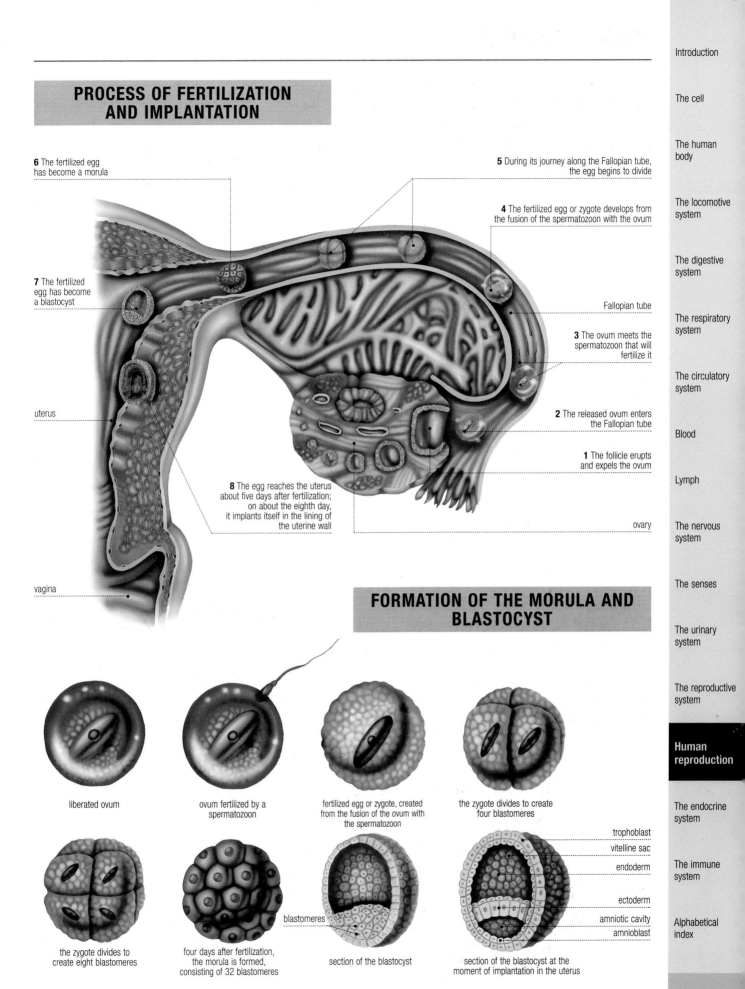

Introduction

The cell

The human body

The locomotive system

The digestive system

The respiratory system

The circulatory system

Blood

Lymph

The nervous system

The senses

The urinary system

The reproductive system

Human reproduction

The endocrine system

The immune system

Alphabetical index

PROCESS OF FERTILIZATION AND IMPLANTATION

6 The fertilized egg has become a morula

7 The fertilized egg has become a blastocyst

uterus

vagina

8 The egg reaches the uterus about five days after fertilization; on about the eighth day, it implants itself in the lining of the uterine wall

5 During its journey along the Fallopian tube, the egg begins to divide

4 The fertilized egg or zygote develops from the fusion of the spermatozoon with the ovum

Fallopian tube

3 The ovum meets the spermatozoon that will fertilize it

2 The released ovum enters the Fallopian tube

1 The follicle erupts and expels the ovum

ovary

FORMATION OF THE MORULA AND BLASTOCYST

liberated ovum

ovum fertilized by a spermatozoon

fertilized egg or zygote, created from the fusion of the ovum with the spermatozoon

the zygote divides to create four blastomeres

the zygote divides to create eight blastomeres

four days after fertilization, the morula is formed, consisting of 32 blastomeres

blastomeres

section of the blastocyst

trophoblast

vitelline sac

endoderm

ectoderm

amniotic cavity

amnioblast

section of the blastocyst at the moment of implantation in the uterus

GESTATION

Gestation, or pregnancy, begins at the moment of fertilization and ends about **nine months** later with the birth of a baby. During this period, the successive divisions of the egg cell give rise to the formation of an **embryo**. At three months, the **embryo** already has a clearly human appearance and is called a **fetus**. It needs only to mature in the mother's uterus for the appropriate length of time before it is ready for its autonomous life.

DEVELOPMENT OF THE EMBRYO

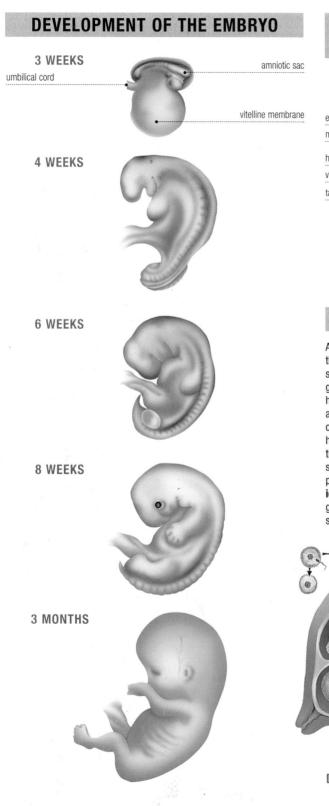

3 WEEKS
umbilical cord
amniotic sac
vitelline membrane

4 WEEKS

6 WEEKS

8 WEEKS

3 MONTHS

ELEMENTS OF THE EMBRYO AT FOUR WEEKS

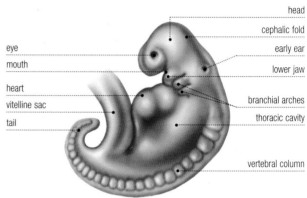

eye
mouth
heart
vitelline sac
tail

head
cephalic fold
early ear
lower jaw
branchial arches
thoracic cavity
vertebral column

TWINS

Although in most pregnancies a single fetus is created, it is possible that two or even more fetuses can develop in the mother's womb simultaneously, each of which produce babies. If two occur, these are generically known as **twins**. Sometimes this is because two different ova have been fertilized by different spermatozoa. In this case, they develop as **dizygotic** or **fraternal twins** and each one has its own placenta. They can be the same sex or different sexes and will be as similar as if they had been born separately. On other occasions, it is possible that a zygote that is created from the fusion of a single ovum with a single spermatozoon might separate into two or more fragments, each of which produces its own embryo. In this case, they develop into **monozygotic** or **identical twins**. These share a single placenta and have the same genetic traits, which is why they are always the same sex and are very similar.

MONOZYGOTIC TWINS

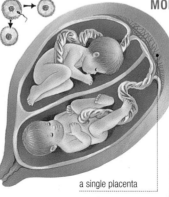

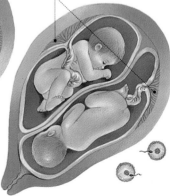

two placentas

a single placenta

DIZYGOTIC TWINS

DEVELOPMENT OF THE FETUS IN THE WOMB

THIRD WEEK

The cells multiply to form all the tissues and organs. The structures that will become all the different organs, skeleton, blood vessels, and nerves appear.

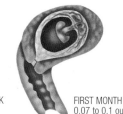

FIRST MONTH
0.07 to 0.1 ounce
(2 to 3 g)/0.3 inch
(0.75 cm)

The heart begins to beat, and the vertebral column and brain appear.

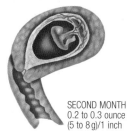

SECOND MONTH
0.2 to 0.3 ounce
(5 to 8 g)/1 inch
(3 cm)

The feet and hands can be seen, and the organs are recognizable. From the second month onward, the fetus develops rapidly.

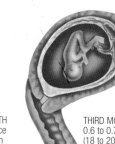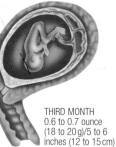

THIRD MONTH
0.6 to 0.7 ounce
(18 to 20 g)/5 to 6
inches (12 to 15 cm)

The fetus begins to look like a human being, with a very large head compared with the rest of the body.

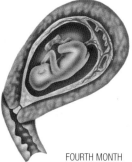

FOURTH MONTH
4 ounces (120 g)/7 to
8 inches (18 to 20 cm)

The digestive tract, liver, pancreas, and kidneys begin to work. Hair and nails appear. The fetus begins to move its arms and legs.

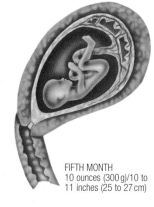

FIFTH MONTH
10 ounces (300 g)/10 to
11 inches (25 to 27 cm)

Maturing of the nervous system. The mother begins to feel the fetus moving. Eyebrows, eyelids, and hair appear.

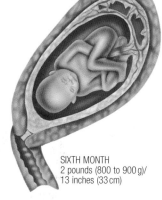

SIXTH MONTH
2 pounds (800 to 900 g)/
13 inches (33 cm)

The bone marrow begins to produce red blood cells. The fetus acquires a pinkish color as the blood capillaries become visible. The lungs mature.

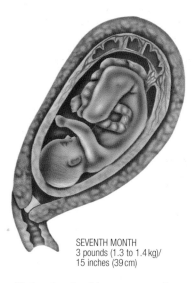

SEVENTH MONTH
3 pounds (1.3 to 1.4 kg)/
15 inches (39 cm)

The lungs have the minimum structure to allow the baby to survive in the event of a premature birth. The fetus has undergone massive growth. The internal organs continue to mature to be ready for life in the outside world.

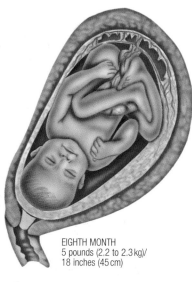

EIGHTH MONTH
5 pounds (2.2 to 2.3 kg)/
18 inches (45 cm)

The lungs are ready to breathe. The skin is smooth and pink in color.

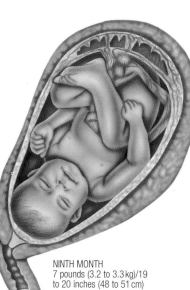

NINTH MONTH
7 pounds (3.2 to 3.3 kg)/19
to 20 inches (48 to 51 cm)

The fetus is fully formed and the thorax is prominent. It positions itself within the mother's pelvis to prepare for the moment of birth. It seems bigger because it has fat under the skin.

Introduction

The cell

The human body

The locomotive system

The digestive system

The respiratory system

The circulatory system

Blood

Lymph

The nervous system

The senses

The urinary system

The reproductive system

Human reproduction

The endocrine system

The immune system

Alphabetical index

PLACENTA

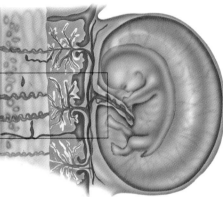

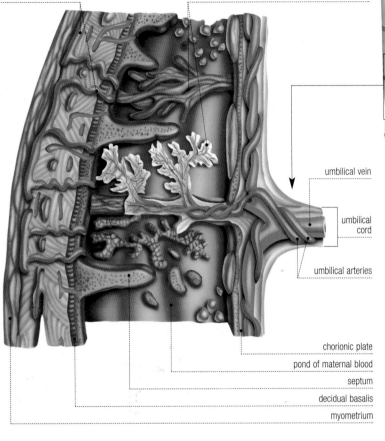

vessels of
maternal blood

chorionic villi

umbilical vein

umbilical
cord

umbilical arteries

chorionic plate

pond of maternal blood

septum

decidual basalis

myometrium

The placenta is an organ that develops during gestation and **forms a bridge** between the organisms of the mother and the fetus. It is formed soon after implantation from the embryo's external tissue, called the **chorion**, and the uterine membrane, or **decidua**, which adapts itself for pregnancy. Maternal blood vessels are connected to the placenta, and from here further blood vessels are connected to the fetus along the **umbilical cord**. An essential **interchange of substances** takes place in the placenta, between the mother's blood and the blood of the fetus, although the blood is not in direct contact. Nutrients and oxygen pass from the maternal circulatory system to the fetal one. Fetal metabolic residues pass in the opposite direction to be eliminated by the mother's body.

DEVELOPMENT OF THE FETUS IN THE WOMB

THIRD MONTH
The fetus is completely formed.
It begins a period of very rapid
growth.

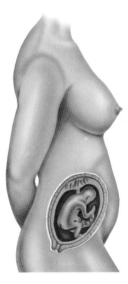

FIFTH MONTH
The fetus begins to move
actively and to react to
sounds.

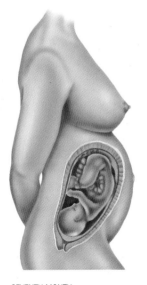

SEVENTH MONTH
Major maturation of the internal
organs. The fetus is capable of
surviving premature birth.

NINTH MONTH
The fetus is fully developed. It fits
perfectly within the mother's
pelvis as it prepares for birth.

ABDOMEN OF A PREGNANT WOMAN

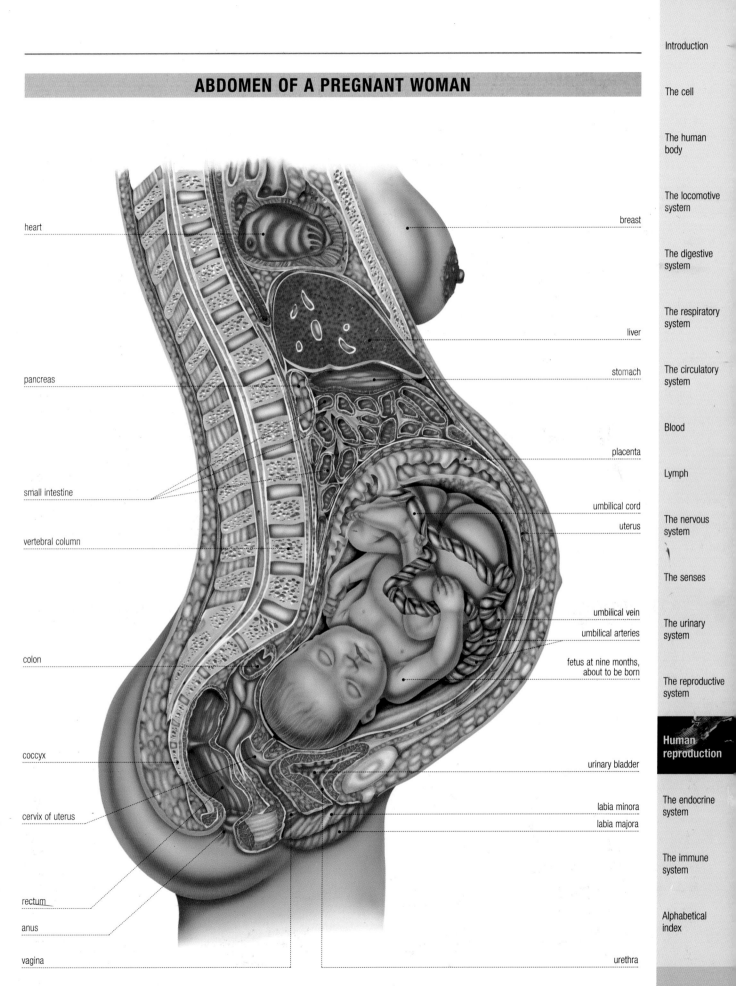

heart

breast

liver

stomach

pancreas

placenta

umbilical cord

uterus

small intestine

vertebral column

umbilical vein

umbilical arteries

colon

fetus at nine months, about to be born

coccyx

urinary bladder

cervix of uterus

labia minora

labia majora

rectum

anus

vagina

urethra

Introduction

The cell

The human body

The locomotive system

The digestive system

The respiratory system

The circulatory system

Blood

Lymph

The nervous system

The senses

The urinary system

The reproductive system

Human reproduction

The endocrine system

The immune system

Alphabetical index

BIRTH

Following the gestation period of nine months, an incomparable event occurs: the **birth** of a baby capable of living an autonomous life outside the mother's womb, although it will still need to be looked after by its parents for a long time to come.

Birth is a lengthy process, divided into **different phases**. During this process, the orifice of the neck of the uterus dilates and the walls of the womb strongly contract to expel first the fetus and then the placenta into the exterior.

POSITIONING OF THE FETUS

During most of the pregnancy, the fetus floats freely in the liquid that surrounds it in the amniotic sac. As it grows, it has less space to move and its movements are restricted. As the moment of birth approaches, the fetus descends, and its head fits into position between the bones of the mother's pelvis. Now everything is ready for the process of birth to begin.

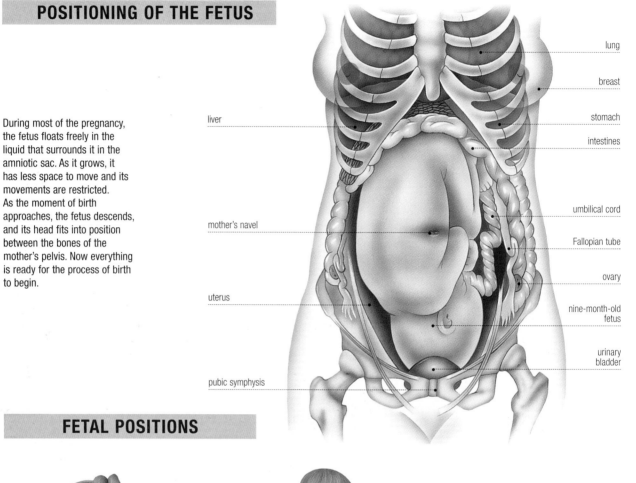

liver

mother's navel

uterus

pubic symphysis

lung

breast

stomach

intestines

umbilical cord

Fallopian tube

ovary

nine-month-old fetus

urinary bladder

FETAL POSITIONS

OCCIPITAL PRESENTATION

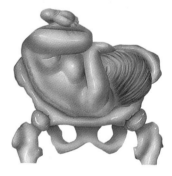

BREECH PRESENTATION

TRANSVERSE PRESENTATION

Under normal conditions, the fetus adopts the typical position for birth, called **occipital presentation**. Its head points downward, its bottom points upward, and its arms and legs are folded. However, sometimes the fetus finds itself in a different position. **Breech presentation** is when it has its head pointing upward. **Transverse presentation** is when it is perpendicular to the mother's pelvis. In these cases, the birth is more difficult. Very often a caesarian section will be used, a surgical intervention in which an incision is made in the mother's uterus in order to extract the baby.

THE DELIVERY PROCESS

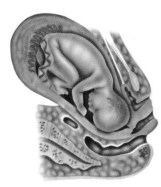

Between three and four weeks (for first-time mothers) and a few hours (for experienced mothers) before birth, the head of the fetus places itself at the entrance to the mother's pelvis.

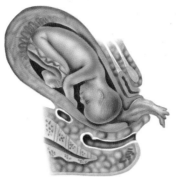

The muscles of the uterus begin to contract at irregular intervals and with irregular degrees of intensity. These contractions cause the bag that surrounds the fetus to break and discharge the liquid it contains (about 2 quarts (2 l)) and pushes the baby toward the outside.

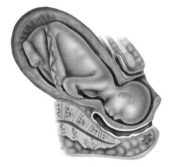

The neck of the uterus begins to dilate to about 4 inches (10 cm) to aid the passage of the fetus to the exterior. The contractions become increasingly intense and frequent.

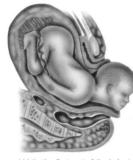

In a normal birth, the first part of the baby to emerge is the head. If the mother has not dilated sufficiently, making an incision in the peroneum is necessary to avoid complications.

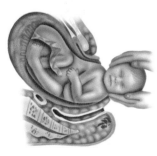

Once the head has emerged, the baby's body turns and begins to emerge. The duration of this phase of the process, as with the other phases, varies.

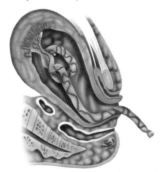

After the baby has emerged from the mother's womb, it is still connected to the placenta by means of the umbilical cord, which has to be cut. The placenta stays inside the womb along with the residues of the birth.

The placenta, the rest of the umbilical cord, and any other residues are expelled about 15 minutes later thanks to strong contractions by the muscles of the uterus.

Once the placenta and umbilical cord have been expelled from the womb, the delivery process has finished.

THE EXPULSION PHASE

The culminating moment in the birth is the expulsion phase of the fetus. The crown of the baby's head appears in the mother's vulva. Shortly afterward, its head emerges into the exterior, followed easily by the rest of the body.

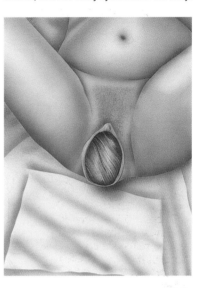

From the moment when the contractions of the uterus start until the fetus finally emerges is usually between 6 and 12 hours for first-time mothers and around 4 hours for mothers who have already had another baby.

Introduction

The cell

The human body

The locomotive system

The digestive system

The respiratory system

The circulatory system

Blood

Lymph

The nervous system

The senses

The urinary system

The reproductive system

Human reproduction

The endocrine system

The immune system

Alphabetical index

THE ENDOCRINE SYSTEM

The endocrine system is made up of a series of internal secretory glands that produce hormones and empty them directly into the bloodstream. Hormones are chemical messages that reach their destination in small quantities by means of the bloodstream. Some hormones act on specific organs, accelerating or inhibiting certain reactions. Others act on all tissues, regulating, among other things, the metabolism as well as growth.

GLANDS OF THE ENDOCRINE SYSTEM

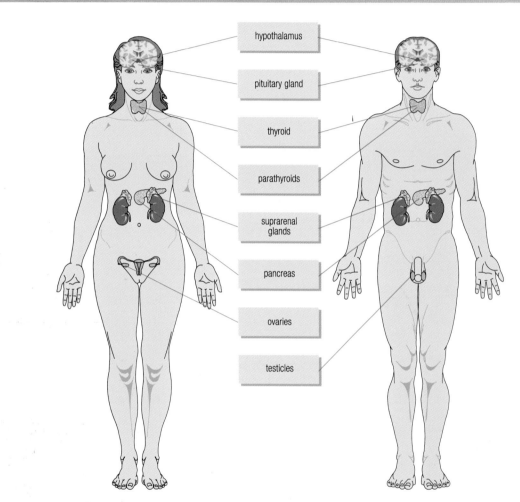

hypothalamus

pituitary gland

thyroid

parathyroids

suprarenal glands

pancreas

ovaries

testicles

HYPOTHALAMUS AND PITUITARY GLAND

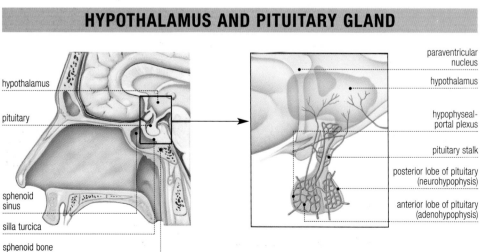

hypothalamus

pituitary

sphenoid sinus

silla turcica

sphenoid bone

paraventricular nucleus

hypothalamus

hypophyseal-portal plexus

pituitary stalk

posterior lobe of pituitary (neurohypophysis)

anterior lobe of pituitary (adenohypophysis)

The hypothalamus and the pituitary gland are two small structures located in the base of the cerebrum that have a particular anatomic relationship. On one hand, some of the neurons of the hypothalamus have extensions that reach the posterior lobe of the pituitary gland (neurohypophysis). On the other hand, a system of blood vessels, the hypophyseal-portal plexus, carries hormonal products produced by the hypothalamus to the anterior lobe of the pituitary gland (adenohypophysis).

FUNCTIONES OF THE HYPOTHALAMUS

The hypothalamus acts as a bridge between the nervous system and the endocrine system. It contains neurological centers that regulate various bodily functions. By means of its hormonal secretions, it modulates the activity of the pituitary gland.

HORMONAL SECRETION OF THE PITUITARY GLAND

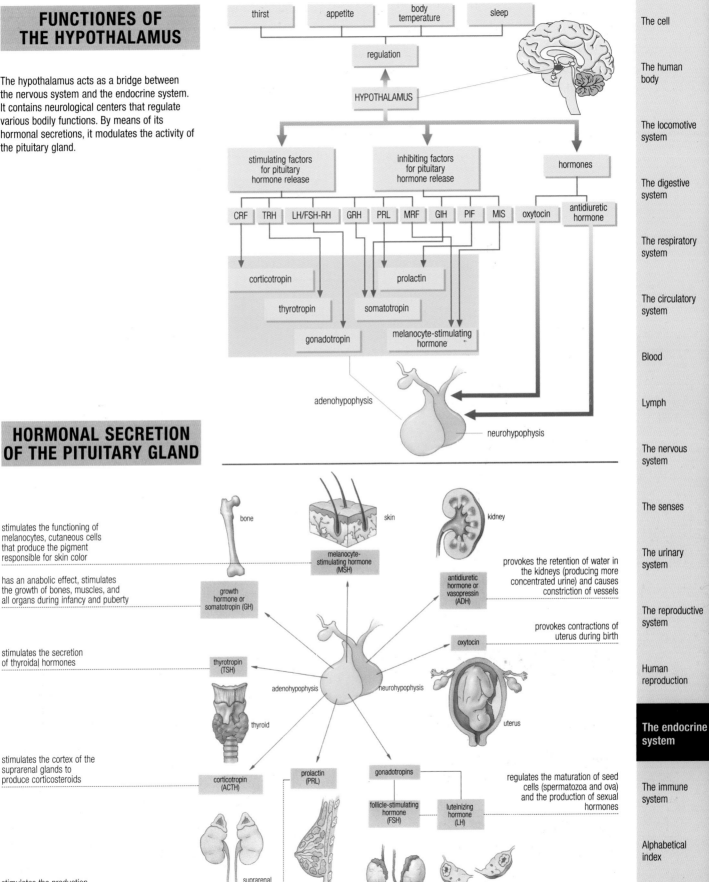

thirst · appetite · body temperature · sleep

regulation

HYPOTHALAMUS

stimulating factors for pituitary hormone release

inhibiting factors for pituitary hormone release

hormones

CRF · TRH · LH/FSH-RH · GRH · PRL · MRF · GIH · PIF · MIS · oxytocin · antidiuretic hormone

corticotropin

thyrotropin

prolactin

somatotropin

gonadotropin

melanocyte-stimulating hormone

adenohypophysis

neurohypophysis

bone

skin

kidney

melanocyte-stimulating hormone (MSH)

stimulates the functioning of melanocytes, cutaneous cells that produce the pigment responsible for skin color

has an anabolic effect, stimulates the growth of bones, muscles, and all organs during infancy and puberty

growth hormone or somatotropin (GH)

antidiuretic hormone or vasopressin (ADH)

provokes the retention of water in the kidneys (producing more concentrated urine) and causes constriction of vessels

oxytocin

provokes contractions of uterus during birth

stimulates the secretion of thyroidal hormones

thyrotropin (TSH)

adenohypophysis

neurohypophysis

thyroid

uterus

stimulates the cortex of the suprarenal glands to produce corticosteroids

corticotropin (ACTH)

prolactin (PRL)

gonadotropins

follicle-stimulating hormone (FSH)

luteinizing hormone (LH)

regulates the maturation of seed cells (spermatozoa and ova) and the production of sexual hormones

suprarenal glands

breasts

testicles

ovaries

stimulates the production of mother's milk

Introduction

The cell

The human body

The locomotive system

The digestive system

The respiratory system

The circulatory system

Blood

Lymph

The nervous system

The senses

The urinary system

The reproductive system

Human reproduction

The endocrine system

The immune system

Alphabetical index

THYROID GLAND

ANTERIOR VIEW

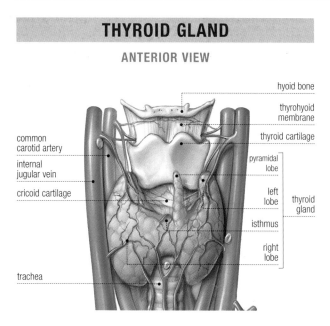

common carotid artery
internal jugular vein
cricoid cartilage
trachea

hyoid bone
thyrohyoid membrane
thyroid cartilage
pyramidal lobe
left lobe
isthmus
right lobe
thyroid gland

The thyroid gland is located in the anterior part of the neck. It consists of two *lateral lobes* that surround the trachea and are linked by a narrow section of tissue called the *isthmus*, although the thyroid sometimes also has a small extension at the top called the *pyramidal lobe*. The thyroid gland produces hormones that stimulate the metabolic reactions of the body, increasing the cellular consumption of oxygen and the production of heat. These are both essential for the physical growth and mental development of children. These hormones are **thyroxin** and **triiodothyronine.**

REGULATION OF THYROID FUNCTION

The production of thyroid hormones depends on stimulation by **thyrotropin hormone** (TSH) produced by the pituitary. Secretion of this hormone depends, in turn, on the **thyrotropin-releasing hormone** (TRH) produced by the hypothalamus. A delicate balance between them allows the body to adapt the levels of thyroid hormones in the bloodstream. The stimulation of the gland increases when there is a deficit (positive feedback) while it decreases if there is an excess (negative feedback).

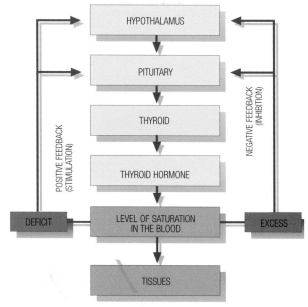

POSITIVE FEEDBACK (STIMULATION)
NEGATIVE FEEDBACK (INHIBITION)

HYPOTHALAMUS
PITUITARY
THYROID
THYROID HORMONE
DEFICIT — LEVEL OF SATURATION IN THE BLOOD — EXCESS
TISSUES

LOCATION OF THE PARATHYROID GLANDS

The parathyroid glands are four very small glands located on the posterior face of the two lateral lobes of the thyroid gland. Their function is to produce **parathyroid hormone**, or **parathormone**, which is involved in the regulation of calcium and phosphorous levels in the blood.

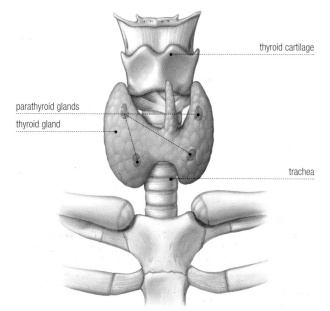

parathyroid glands
thyroid gland

thyroid cartilage

trachea

PARATHYROID HORMONE ACTIVITY

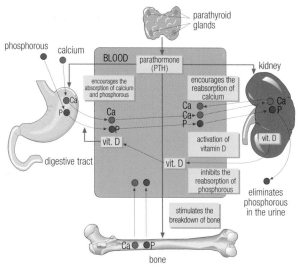

parathyroid glands
phosphorous   calcium
kidney
BLOOD   parathormone (PTH)
encourages the absorption of calcium and phosphorous
encourages the reabsorption of calcium
Ca
Ca
P
Ca
Ca
P
Ca
P
vit. D
vit. D
activation of vitamin D
vit. D
digestive tract
inhibits the reabsorption of phosphorous
eliminates phosphorous in the urine
stimulates the breakdown of bone
Ca   P
bone

The basic purpose of parathyroid hormone is to increase levels of calcium in the blood. To do this, it acts on different levels. It encourages the absorption of this mineral along the digestive tract, it breaks down bone tissue in order to release calcium deposits, and it reduces loss of calcium in the urine.

LOCATION OF THE SUPRARENAL GLANDS

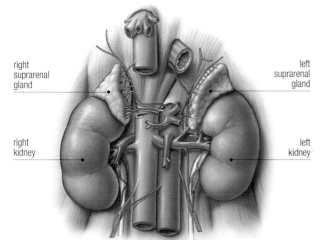

right suprarenal gland

left suprarenal gland

right kidney

left kidney

The suprarenal, or adrenal, glands are two pyramid-shaped glands located, like caps, above the upper tips of the kidneys. Inside are two distinct parts, which have different compositions and different activities:

• the central part, the **suprarenal medulla**, consists of nerve tissue that specializes in the production of catecholamines, such as adrenaline and noradrenaline;

• the outer part, the **suprarenal cortex**, operates under the stimulus of adrenocorticotropin (ACTH) from the pituitary gland and consists of three layers of glandular tissue that produce different corticosteroid hormones: the **zona reticularis** produces androgens, such as dehydroepiandrosterone, which are male sex hormones; the **zona fasciculata** produces glucocorticoids; and the **zona glomerulosa** secretes mineralocorticoids.

SECTION OF A SUPRARENAL GLAND

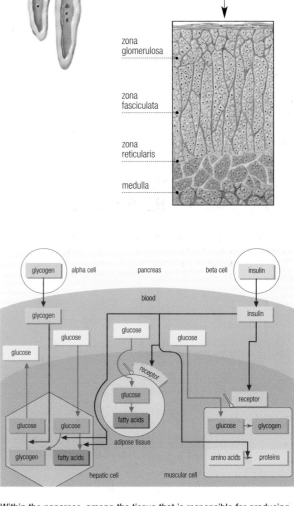

cortex

medulla

zona glomerulosa

zona fasciculata

zona reticularis

medulla

ENDOCRINE TISSUE OF THE PANCREAS

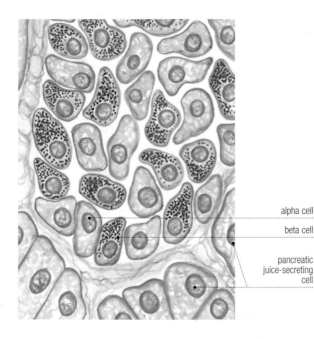

alpha cell

beta cell

pancreatic juice-secreting cell

Within the pancreas, among the tissue that is responsible for producing digestive secretions, are found groups of cells called the **islets** or **islands of Langerhans**. These islets are made up of two types of cells responsible for secreting and emptying hormones directly into the blood to regulate the metabolism of glucose and the concentration of glucose in the blood: **alpha cells** fabricate glycogen, and **beta cells** produce insulin.

Introduction

The cell

The human body

The locomotive system

The digestive system

The respiratory system

The circulatory system

Blood

Lymph

The nervous system

The senses

The urinary system

The reproductive system

Human reproduction

The endocrine system

The immune system

Alphabetical index

THE IMMUNE SYSTEM

ORGANS OF THE IMMUNE SYSTEM

The immune system is responsible for **defending the body** against the many germs that surround us, which represent a multitude of tiny, and potentially dangerous, foreign elements. To do this, the body depends on the activity of **white blood cells**, or leukocytes, which are produced by different organs of the body and which travel around it constantly, looking for all kinds of harmful agents in order to **destroy** or **deactivate** them.

thymus
small lymphoid organ where white blood cells, called lymphocyte T cells, mature during the fetal period and childhood in order to be able to perform their particular function

lymphatic ganglia
node formations interspersed along the lymphatic ducts; they store numerous white blood cells and act as a filter for germs and impurities

spleen
organ that fabricates some white blood cells and acts as a filter for germs and impurities in the blood that circulate through it

bone marrow
tissue found inside various bones of the skeleton, responsible for producing blood cells, including white blood cells.

LOCATION OF THE THYMUS

The thymus is located in the center of the thorax, behind the sternum. Its anatomic characteristics undergo an interesting evolution over the course of its life. It is a very important organ during childhood because that is where lymphocyte T cells, a type of white blood cell, mature. For this reason, during childhood, the thymus can grow to weigh around 1.5 ounces (45 g). However, its activity decreases after puberty and eventually almost ceases altogether, causing the organ to atrophy progressively, so that in adults it weighs a maximum of about 0.5 ounce (15 g).

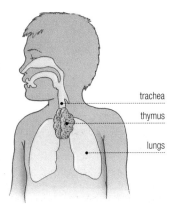

trachea

thymus

lungs

WHITE BLOOD CELLS

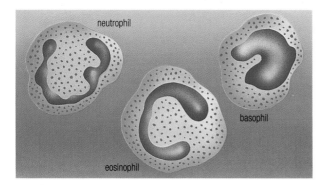

GRANULOCYTES

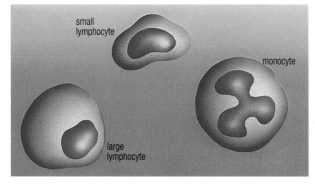

AGRANULOCYTES

TYPES OF WHITE BLOOD CELL

Type	Percentage of total	Function
Granulocytes neutrophils	45 to 75 percent	phagocytosis
Granulocytes eosinophils	1 to 3 percent	involved in allergic reactions and in defense against parasites
Granulocytes basophils	1 percent	involved in allergic reactions
Monocytes	3 to 7 percent	phagocytosis
Lymphocytes	25 to 30 percent	T lymphocytes: coordinating immune reaction and immune cell response B lymphocytes: humoral immune reaction

MECHANISM OF NONSPECIFIC IMMUNITY

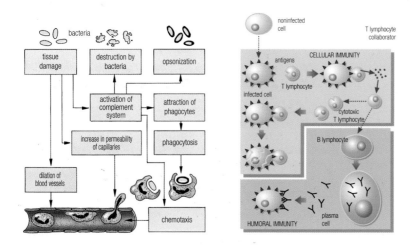

If a foreign agent invades the organism, the immune system first releases a nonspecific response based on mechanisms already present from birth. This is the innate immune system.

If this is not sufficient, a specific response to the aggressor is produced, either by the direct action of white blood cells or by means of antibodies that some of them produce (plasma cells). This is called acquired immunity, which develops over the course of the organism's life as it comes across different germs.

ACTIVE IMMUNIZATION: VACCINATION

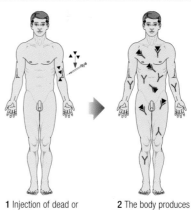

1 Injection of dead or deactivated germs

2 The body produces antibodies against the germ

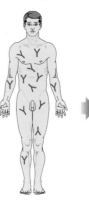

3 The body is now sensitive to the germ

4 In the event of new contact with the germ, the immune system will act immediately

Introduction

The cell

The human body

The locomotive system

The digestive system

The respiratory system

The circulatory system

Blood

Lymph

The nervous system

The senses

The urinary system

The reproductive system

Human reproduction

The endocrine system

The immune system

Alphabetical index

A

abdomen (location of) 12–13
abdomen (of pregnant woman) 83
abdominal lymph nodes 51
abductor digiti minimi muscle 28
abductor peroneus longus muscle 29
abductor pollicis muscle 28
acetabulum (foot) 21
Achilles tendon 26
acinar cells 35
acinar clusters 35
acini (mammary gland) 76
aconeus muscle 26
acromion 20
actin 23
Adam's apple 40
adductor brevis muscle 24–25
adductor digiti minimi muscle 28
adductor hallucis muscle 29
adductor longus muscle 24–25
adductor magnus muscle 24–25
adductor maximus muscle 29
adductor pollicis muscle 28
adenine 11
adenohypophysis 86
adipose tissue 66
adrenal glands 89
amniotic sac 80
anal duct 30
anal sphincter 37
antebrachium 20
anterior brachii muscle 24–25
anterior cardiac veins 45
anterior jugular vein 47
anterior labyrinth (ear) 62
anterior papillary muscle 44
anterior tibial artery 46
anterior tibial vein 47
antidiuretic hormone (vasopressin) 87
anus 30, 72, 75
anus (of pregnant woman) 83
anvil (incus) 62
aorta 32, 42–44
aortic valve 44
aqueous humor 58
arachnoid 55
arch of aorta 46
areola 76
arm (location of) 12–13
armpit (location of) 12–13
arteries 42–46
arteries, principal 46
arterioles 42
artery, section 46
articular capsule 23
articular cartilage 23
articular cavity 23
articular liquid 23
ascending aorta 46
ascending colon 30
atlas 19, 22
atrioventricular valve 44
atrium 43–44
auditory canal 62
auditory or vestibulocochlear nerve (VIII cranial
   pair) 53
auditory ossicles 62
auricularis posterior muscle 26–27
auricularis superior muscle 26

axillary artery 46
axillary lymph nodes 51
axis 19, 22
axon 52
azygus vein 47

B

back (location of) 12–13
basal membrane (artery) 46
basal membrane (vein) 47
basilic vein 47
biceps 28
biceps brachii muscle 24–25
biceps femoris muscle 26, 29
biceps muscle 24
bile 36
bile duct 35–36
Billroth's cords 49
bipenniform muscle 24
birth 84
blastocyst 79
blastomere 79
blood 48
blood cells 48
blood vessel 14, 42
bolus of food 30
bone, formation and growth 14
bone fractures 15
bone marrow 49, 90
bone types 15
Bowman's capsule 70
Bowman's membrane 60
brachial veins 47
brachialis muscle 26
brachiocephalic trunk 46
brachioradialis muscle 25–26
brachium 20
brain 52
brain stem 53
breast 13, 76
breast (location of) 12–13
breast (of pregnant woman) 83
breech presentation of fetus 84
bronchi 38, 41
bronchial tree 41
buccal cavity 30–31
buccinator muscle 27
bundle of His 45
buttock (location of) 12–13

C

calcaneus (foot) 17, 21
calyx 69
canaliculi 14, 35
cancellous bone 14–15
canines 31
capillary 42
capillary, section 46
capitate bone 20
cardia 33
cardiac cycle 45
cardiac valves 44
cardiac veins 45
carotid artery 88
carpal bones 16–17
carpus 20
cartilage 14
catecholamines 89
caudal nucleus 54
cecum 37
celiac trunk 46

cell division 11
cellular membrane or cytoplasm 10
cellular organelles 10
cementum 31
centrioles 10
centromere 11
cephalic vein 47
cerebellum 53–55
cerebral ventricles 55
cerebrospinal fluid 55
cerebrum 53–55
cervical roots 53
cervical vertebrae 19
cervix of uterus (of pregnant woman) 83
cheek (location of) 12–13
chin (location of) 12–13
choanae 39
chorion 82
choroid 58
chromatin 11
chromosome 11
ciliary body 58
circular muscle 24
circulatory apparatus 42
circumflex branch of left coronary artery 45
clavicle 16–17, 20, 22
clitoris 74
coccyx 17, 19
coccyx (of pregnant woman) 83
cochlea 62–63
coitus 78
colon 37
colon (of pregnant woman) 83
columns of Bertin 69
common bile duct 36
common left carotid artery 46
common ocular motor nerve
   (III cranial pair) 53
common right carotid artery 46
compact bone 14–15, 49
concha 18
condylarthrosis 22
cones in retina 60
conjunctiva 58, 60
connective tissue 31
coracoid process (scapula) 20
cornea 58, 60
coronary artery 45
coronary blood vessels 45
coronary sinus 45
coronary system of electric impulses 45
corpus albicans 76
corpus callosum 54
corpus cavernosum 73
corpus luteum 76
corpus spongiosum 73
corticosteroids 87
corticotropin (hormone) 87
costal process (vertebra) 19
cotyloid cavity 23
cotyloid cavity (foot) 21
cotyloid flange 23
Cowper's gland 71
cricoid cartilage 32, 40, 88
crown 31
cubital posterior muscle 26
cuboid bone (foot) 21
cytoplasm 10
cytosine 11

D

decidua 82
deciduous teeth 31
deep femoral vein 47
defects of sight 60
delivery process 85
deltoid muscles 25–26, 28
dendrites 52
dentin 31
deoxyribose 11
depressor anguli oris muscle 27
depressor labii inferioris muscle 27
depressor septi nasi muscle 27
dermis 66
Descemet's membrane 60
descending aorta 46
descending colon 30
diaphragm 32, 36, 38
diaphysis 15
diarthrosis 22
diastole 44–45
digestive system 30
digiti minimi tendon 28
digitus profundus tendon 28
dislocations 22
distal interphalangeal joint 28
distal phalanx 20
dizygotic twins 80
DNA 11
dorsal artery of foot 46
dorsal interosseous muscles 28–29
duct of Santorini 35
duct of Wirsung 35
ductus deferens 72
duodenum 30, 33–35
dura mater 55

E

ear 12–13, 62
eardrum 62
ejaculatory duct 72
elbow 12–13, 20
embryo 80
enamel 31
enarthrosis 22
endocrine system 86
endometrium 77
endomysium 23–24
endoplasmic reticulum 10
endosteum 15
endothelium (artery) 46
endothelium (vein) 47
epicardium 44
epidermis 66
epididymis 72–73
epidural space 55
epiglottis 39–40, 65
epimysium 23–24
epiphysis 15
epitrochlear lymph nodes 51
equilibrium 62
erythrocytes 48
esophagus 30–33, 39
ethmoid bone 18
ethmoidal sinus 39
eukaryotic cells 11
eustachian tube 39, 62
extensor carpi radialis tendons 28
extensor carpi ulnaris brevis muscle 26
extensor carpi ulnaris longus muscle 26

extensor digiti minimi muscle 28
extensor digitorum brevis muscle 29
extensor digitorum longus muscle 25
extensor digitorum longus tendon 29
extensor digitorum muscle 26
extensor hallucis longus muscle 29
extensor hallucis longus tendon 29
extensor indicis muscle 26
extensor pollicis brevis muscle 28
extensor pollicis brevis tendon 28
extensor pollicis longus muscle 26, 28
extensor pollicis longus tendon 28
external abdominal oblique muscle 25–26
external capsule 54
external carotid artery 46
external elastic membrane (artery) 46
external iliac artery 46
external iliac vein 47
external jugular vein 47
external ocular nerve (VI cranial pair) 53
eye (location of) 12–13
eyeball 59

F

face (location of) 12–13
facial muscles 27
facial nerve (VII cranial pair) 53
Fallopian tube 74–75, 79
feces 37
femoral artery 46
femoral vein 47
femur 16–17, 21–23
fertilization 78
fetus 80, 82
fetus position 84
fibula 16, 21–22
finger extensors 28
finger flexors 28
fingers 12–13, 20
fissure of Rolando 54
flat muscle 24
flexor carpi ulnaris muscle 25
flexor digiti minimi muscle 28
flexor digitorum longus muscle 26
flexor longus pollicis muscle 26
flexor pollicis muscle 28
follicle cells 76
follicle-stimulating hormone 87
foot 12–13, 21
forearm (location of) 12–13
forehead (location of) 12–13
frontal bone 16
frontal sinus 39
frontalis muscle 25, 27

G

gallbladder 30, 36
gastric juice 33
gastrocnemius muscle 25–26, 29
genital organs (female) 74
genital organs (male) 72
gestation 80
gingival epithelium 31
glans 72
glenoid cavity 20
glomerule (kidney) 70
glossopharyngeal nerve (IX cranial pair) 53
gluteus maximus muscle 26, 29
gluteus medius 25–26
glycogen 30, 89
golgi apparatus 10

gonadotropins (hormone) 87
gracilis muscle 26
gray matter 56
groin 12–13
groin (location of) 12–13
growth hormone (somatotropin) 87
growth plate 14
guanine 11
gustatory pore 65
gyrus 54

H

hair (location of) 12–13
hair follicle 67
hamate bone 20
hammer (malleus) 62
hand 12–13, 20, 28
hard palette 31
haustra 37
head (location of) 12–13
hearing 62
heart 42–45
Heister's duct 36
hematopoiesis 48
Henle's loop 70
hilum of kidney 69
hilum of spleen 49
hip bone 21, 23
hip (location of) 12–13
hormones 86
humerus 15–16, 20–22
hydrochloric acid 33
hydrogen bridges 11
hymen 74
hyoid bone 40, 88
hypermetropia (farsightedness) 60
hypodermis 66
hypoglossal nerve (XII cranial pair) 53
hypothalamus 86–88

I

ileocecal valve 34
ileum 30, 34, 37
iliac tuberosity (foot) 21
iliacus muscle 24–25
iliocostalis muscle 26
iliofemoral ligament 23
iliotibial tract (muscle) 26
ilium 16–17, 21–22
immune system 90
incisors 31
incus 18, 62
index finger 20
inferior maxillary nerve 53
inferior mesenteric artery 46
inferior mesenteric vein 47
inferior vena cava 44
infraspinatus fossa (scapula) 20
infraspinatus muscle 26
inguinal lymph nodes 51
insulin 35, 89
intermediary nerve 53
intermediate cuneiform bone (foot) 21
intermediate vein 47
internal abdominal oblique muscle 26
internal capsule 54
internal carotid artery 46
internal elastic membrane (artery) 46
internal iliac artery 46
internal iliac vein 47
internal jugular vein 47

Introduction

The cell

The human body

The locomotive system

The digestive system

The respiratory system

The circulatory system

Blood

Lymph

The nervous system

The senses

The urinary system

The reproductive system

Human reproduction

The endocrine system

The immune system

Alphabetical index

internal saphenous vein 47
internal thoracic vein 47
interventricular septum 44
iris 58–59
ischial tuberosity (foot) 21
ischium bone 16, 21
islands of Langerhans 89
isthmus (thyroid gland) 88

J
jejunum 30, 34
joints 22

K
kidney 68–70
knee 12–13, 21
knee (location of) 12–13
Krause's corpuscle 66

L
labium majora 74–75
labium minora 74–75
labyrinth (ear) 63
lacrimal apparatus 59
lactiferous ducts 76
lactiferous sinuses 76
lacunae 14
lamina of elastic fibers (artery) 46
large intestine 34, 37
larynx 38–40
lateral cuneiform bone (foot) 21
lateral head of gastrocnemius muscle 26
lateral peroneus brevis tendon 29
lateral peroneus longus tendon 29
latissimus dorsii muscle 25–26
left atrioventricular valve 44
left atrium 44
left common brachiocephalic vein 47
left common iliac artery 46
left common iliac vein 47
left coronary artery 45
left pulmonary vein 44
left subclavian vein 47
left ventricle 44
leg 12–13, 21
lens 58
lenticular nucleus 54
leukocytes 48, 90
levator anguli oris muscle 27
levator labii superioris alaeque nasi muscle 27
levator labii superioris muscle 27
ligaments 23
link between blood and lymph in abdominal
    circulation 50
link between blood and lymph in pulmonary
    circulation 50
little finger 20
liver 30, 36
long muscle 24
lower dental arch 31
lower lip 31
lumbar vertebrae 19
lumbrical muscles 28
lunate bone 20
lungs 38, 41
luteinizing hormone 87
luxations 22
lymph 50–51
lymph nodes 50–51
lymphatic capillaries 50
lymphatic duct 51
lymphatic ganglia 90

lymphatic system 50
lymphatic vessels 50
lymphocyte T cells 90
lysosome 10

M
macula lutea 58
major cardiac vein 45
malar 18
malleus 18, 62
malpighian cell 49
mamillary body 53
mammary glands 76
mandible 16–18
masseter muscle 25, 27
masticatory muscles 27
maxilla 16–18
maxillary bone 31
maxillary sinus 39
medial cuneiform bone (foot) 21
medial head of gastrocnemius muscle 26
mediastinal lymph nodes 51
mediastinum 43
medulla oblongata 53–55
medullary cavity 14–15
Meissner's corpuscle 66
melanocyte-stimulating hormone 87
meninges 55
meniscus 23
menstrual cycle 77
menstruation 77
mentalis muscle 27
metacarpal bones 16–17, 20
metacarpus 20
metatarsal bones 21
metatarsus 21
microfilament 10
microtubules 10
microvilli 10
middle cardiac vein 45
middle finger 20
middle phalanx 20
minor cardiac vein 45
mitochondrion 10
mitral valve 44
molars 31
monozygotic twins 80
Monro's sulcus 55
Montgomery's tubercles 76
Morgani's ventricles 40
morula 79
mouth 12–13, 30, 38
muscle fiber 23
muscles of the foot 29
muscles of the hand 28
muscles of the lower limb 29
myelin sheath 52
myocardium 44
myofibrils 23
myometrium 82
myopia (nearsightedness) 60
myosin 23

N
nail 67
nasal bone 16–18
nasal concha 39
navel (location of) 12–13
navicular bone (foot) 21
neck (location of) 12–13
neck lymph nodes 51

neck of tooth 31
nephron 70
nervous system 52
network of Purkinje 45
neurohypophysis 87
neuron 52, 67
nipple 13, 76
nipple (location of) 12–13
Nissi bodies 52
nodes of Ranvier 52
nonspecific immunity 91
noradrenaline 89
nose 38
nose (location of) 12–13
nuclear membrane 10–11
nuclear pore 11
nucleolus 10
nucleus 10

O
occipital bone 17
occipital lobe of cerebrum 53
occipital presentation of fetus 84
occipitalis muscle 26–27
odontoid process (vertebra) 19
olfactory bulb 53, 64
olfactory cell 64
olfactory membrane 39, 64
olfactory nerve 53
olfactory tract 53
oocyte 76–77
ophthalmic groove 53
ophthalmic nerve 53
opponens digiti minimi muscle 28
opponens pollicis muscle 28
optic disk 58
optic nerve 58, 60
optic nerve (II cranial pair) 53
orbicularis oculi muscle 27
orbicularis orbis muscle 25
orbicularis oris muscle 27, 25
organ of Corti 62
os coxae 21
osseous matrix 14
osteoblasts 14
osteoclast 14
osteocytes 14
oval (vestibular) window 63
ovarian artery 46
ovarian follicle 76
ovarian vein 76
ovaries 86
ovary 74–79
ovulation 76–77
ovum 76–78
oxytocin (hormone) 87

P
palm (location of) 12–13
palmar interosseous muscles 28
palmaris brevis muscle 25
palmaris longus muscle 25
pancreas 30, 35, 86, 89
papilla (kidney) 69
papillae (taste) 65
paranasal sinuses 39
parathormone 88
parathyroid glands 88
parathyroid hormone 88
parathyroid hormone activity 88
parenchyma 49

parietal bone 16–18
patella 16, 21, 23
pectineus 24
pectineus muscle 25
pectoralis major muscle 25
pelvis or pubis (location of) 12–13
penis 12–13, 72–73
pepsin 33
pericardium 43
perimysium 23–24
periosteum 14–15
peripheral nervous system 52, 57
permanent teeth 31
peroneal artery 46
peroneus anterior muscle 25
peroneus brevis muscle 25
peroneus longus muscle 25
peroneus tendon 29
phalanges 16–17, 21
pharynx 30–31, 38–39
phosphate 11
pia mater 55
pinna 62
pisiform bone 20
pituitary gland 53, 86–88
placenta 82
placenta (of pregnant woman) 83
plantaris muscle 26
plasma 48
platelets 48
platysma 27
platysma muscle 27
polygastric muscle 24
pons 53
popliteal artery 46
popliteal lymph nodes 51
popliteal vein 47
portal vein 47
posterior labyrinth (ear) 62
posterior tibial artery 46
posterior tubercle (vertebra) 19
posterior vein of the left ventricle 45
pregnancy 80
premolars 31
presbyopia (tired vision) 60
procerus muscle 25
prolactin (hormone) 87
pronator muscle 25
prostate gland 73
proximal interphalangeal joint 28
proximal phalanx 20
psoas muscle 25
pubis 16
pubofemoral ligament 23
pulmonary artery 41–43
pulmonary valve 44
pulmonary vein 41–42
pulp 31
putamen 54
pylorus 33–34
pyramidalis nasi muscle 27
pyramids of Malpighi 69
R
radial artery 46
radius 16–17, 20–22
rectum 30, 37
rectum (of pregnant woman) 83
rectus abdominis muscle 25
rectus femoris muscle 25, 29

red pulp (spleen) 49
reflex action 67
Reissner's membrane 63
relationship between the lymphatic system and the
    circulatory system 50
renal artery 46
renal capsule 69
renal circulation 68
renal cortex 69
renal vein 47
reproductive apparatus (female) 74
reproductive apparatus (male) 72
respiratory system 38
retina 58
rhomboid muscle 26
rhomboids 28
ribosome 10
ribs 16–17
right atrioventricular valve or tricuspid valve 44
right atrium 44–45
right brachiocephalic vein trunk 47
right common iliac artery 46
right common iliac vein 47
right coronary artery 45
right marginal artery 45
right subclavian artery 46
right subclavian vein 47
right ventricle 44
ring finger 20
risorius muscle 27
rods in retina 60
round (cochlear) window 63
S
sacrum 16–17, 19
saphenous vein 47
sarcomere 23
sartorius muscle 25, 29
scaphoid bone 20
scapula 16–17, 22–23
Schwann cells 52
sclera 58
scrotum 12–13, 72
sebaceous gland 66
section of a tooth (molar) 31
semen 73
semicircular canal (ear) 62–63
semimembranous muscle 26
seminal vesicle 72
semispinalis capitis muscle 27
semitendinosus muscle 26
serous capsule 49
serratus anterior muscle 25
shin 12–13
short muscle 24
shoulder 12–13
sight 58–60
sigmoid colon 30
sinoatrial node 45
skeleton 16
skull bones 18
small intestine 30, 34
small intestine (of pregnant woman) 83
smell 64
soft palette 31
soleus muscle 26, 29
somatropin (growth hormone) 87
spermatozoon 78
sphenoid bone 16–18

sphenoidal sinus 39
spinal cord 52–55
spinal ganglion 56
spinal nerve (XI cranial pair) 53
spinous process (vertebra) 19
spleen 49, 90
splenic artery 49
splenic vein 47
splenius capitis muscle 27
splenius muscle 26
stapes 18, 25, 62
sternocleidomastoid muscle 25–27
sternum 16
stirrup (stapes) 62
stomach 30–34
sulcus 54
superior maxillary nerve 53
superior mesenteric artery 46
superior mesenteric vein 47
superior vena cava 44, 47
supinator muscle 25
suprahepatic vein 47
suprarenal glands 69, 86–89
swallowing 32
sweat gland 66
symphysis 22
synarthrosis 22
synovial membrane 23
systole 44–45
T
talus 16–17, 21
tarsus 21
taste 65
taste buds 65
taste corpuscles 65
temporal bone 16–18
temporalis muscle 25, 27
tendon cords 44
tensor fasciae latae muscle 25
teres major muscle 26
testicle 72–73, 86
testicular artery 46
testicular vein 47
thalamus 54
thigh 12–13, 21
thoracic duct 51
thoracic vertebrae 19
thorax (location of) 12–13
thrombocytes 48
thumb 12–13, 20
thymine 11
thymus 90
thyrohyoid membrane 88
thyroid 87–88
thyroid cartilage 32, 40, 88
thyroid gland 88
thyroid hormone 88
thyrotropin (hormone) 87
thyrotropin-releasing hormone 88
thyroxin 88
tibia 16–17, 21–23
tibialis anterior muscle 25, 29
tibialis anterior tendon 29
toes 12–13, 21
tongue 30–31, 65
tonsils 31, 39, 65
touch 66
trabeculae 14, 49
trabecular arteries 49

Introduction

The cell

The human
body

The locomotive
system

The digestive
system

The respiratory
system

The circulatory
system

Blood

Lymph

The nervous
system

The senses

The urinary
system

The reproductive
system

Human
reproduction

The endocrine
system

The immune
system

Alphabetical
index

trabecular veins 49
trachea 32, 38–41, 88
transversal elastic muscle fibers (artery) 46
transverse colon 30
transverse presentation of fetus 84
transverse process (vertebra) 19
tranversalis nasi muscle 27
tranvserse carpal tendons 28
trapezium bone 20
trapezius muscle 25–27
trapezoid bone 20
triangle of neck (muscle) 25–26
triceps 28
triceps brachii muscle 25–26
tricuspid valve 44
trigeminal nerve (V cranial pair) 53
triiodothyronine 88
triquetral bone 20
trochlear nerve (IV cranial pair) 53
trochlearthrosis 22
tuberosity of calcaneus (foot) 21
tunica adventitia (artery) 46
tunica adventitia (vein) 47
tunica albuginea 73
tunica intima (artery) 46
tunica intima (vein) 47
tunica media (artery) 46
tunica media (vein) 47
twins 80

tympanic cavity 62
tympanic membrane 62
tympanum 62
U
ulna 16–17, 20
ulnar artery 46
umbilical cord 80, 82
umbilical cord (of pregnant woman) 83
upper dental arch 31
ureter 68
urethra 68, 72
urethra (of pregnant woman) 83
urinary bladder 68, 71
urinary bladder (of pregnant woman) 83
urinary system 68
uterus 74, 77
uterus (of pregnant woman) 83
uvula 31
V
vacuole 10
vagina 74, 75
vagina (of pregnant woman) 83
vagus nerve (X cranial pair) 53
vastus lateralis muscle 25–26, 29
vastus medialis muscle 25, 29
Vater's duct 34
Vater-Pacini corpuscle 66
vein (section) 47
veins 42

veins of the body 47
veins, principal 47
vena cava 42–44
venous valve (vein) 47
ventricle 43–44
ventricular diastole 44
ventricular systole 44
venules 42
vermiform appendix 30
vertebrae 16–19
vertebral body (vertebra) 19
vertebral column 19
visual pathways 60
vitreous humor 58
vocal chords 39–40
vomer or nasal septum 18
vulva 74
W
white blood cells 90–91
white corpuscles 90
white matter 54, 56
white pulp (spleen) 49
womb 82
wrist 12–13
Z
zygomatic bone 18
zygomaticus major muscle 25, 27
zygomaticus minor muscle 27
zygote 78